Physics: A Very Short Introduction

VERY SHORT INTRODUCTIONS are for anyone wanting a stimulating and accessible way into a new subject. They are written by experts, and have been translated into more than 45 different languages.

The series began in 1995, and now covers a wide variety of topics in every discipline. The VSI library currently contains over 600 volumes—a Very Short Introduction to everything from Psychology and Philosophy of Science to American History and Relativity—and continues to grow in every subject area.

Very Short Introductions available now:

ABOLITIONISM Richard S. Newman
ACCOUNTING Christopher Nobes
ADAM SMITH Christopher J. Berry
ADOLESCENCE Peter K. Smith
ADVERTISING Winston Fletcher
AFRICAN AMERICAN RELIGION
  Eddie S. Glaude Jr
AFRICAN HISTORY John Parker
  and Richard Rathbone
AFRICAN POLITICS Ian Taylor
AFRICAN RELIGIONS
  Jacob K. Olupona
AGEING Nancy A. Pachana
AGNOSTICISM Robin Le Poidevin
AGRICULTURE Paul Brassley
  and Richard Soffe
ALEXANDER THE GREAT
  Hugh Bowden
ALGEBRA Peter M. Higgins
AMERICAN CULTURAL HISTORY
  Eric Avila
AMERICAN FOREIGN RELATIONS
  Andrew Preston
AMERICAN HISTORY Paul S. Boyer
AMERICAN IMMIGRATION
  David A. Gerber
AMERICAN LEGAL HISTORY
  G. Edward White
AMERICAN NAVAL HISTORY
  Craig L. Symonds
AMERICAN POLITICAL HISTORY
  Donald Critchlow
AMERICAN POLITICAL PARTIES
  AND ELECTIONS
  L. Sandy Maisel

AMERICAN POLITICS
  Richard M. Valelly
THE AMERICAN PRESIDENCY
  Charles O. Jones
THE AMERICAN REVOLUTION
  Robert J. Allison
AMERICAN SLAVERY
  Heather Andrea Williams
THE AMERICAN WEST Stephen Aron
AMERICAN WOMEN'S HISTORY
  Susan Ware
ANAESTHESIA Aidan O'Donnell
ANALYTIC PHILOSOPHY
  Michael Beaney
ANARCHISM Colin Ward
ANCIENT ASSYRIA Karen Radner
ANCIENT EGYPT Ian Shaw
ANCIENT EGYPTIAN ART AND
  ARCHITECTURE Christina Riggs
ANCIENT GREECE Paul Cartledge
THE ANCIENT NEAR EAST
  Amanda H. Podany
ANCIENT PHILOSOPHY Julia Annas
ANCIENT WARFARE
  Harry Sidebottom
ANGELS David Albert Jones
ANGLICANISM Mark Chapman
THE ANGLO-SAXON AGE John Blair
ANIMAL BEHAVIOUR
  Tristram D. Wyatt
THE ANIMAL KINGDOM
  Peter Holland
ANIMAL RIGHTS David DeGrazia
THE ANTARCTIC Klaus Dodds
ANTHROPOCENE Erle C. Ellis

Available soon:

For more information visit our website

www.oup.com/vsi/

Sidney Perkowitz

PHYSICS

A Very Short Introduction

OXFORD
UNIVERSITY PRESS

OXFORD
UNIVERSITY PRESS

Great Clarendon Street, Oxford, OX2 6DP,
United Kingdom

Oxford University Press is a department of the University of Oxford.
It furthers the University's objective of excellence in research, scholarship,
and education by publishing worldwide. Oxford is a registered trade mark of
Oxford University Press in the UK and in certain other countries

First edition published in 2019

Published in the United States of America by Oxford University Press
198 Madison Avenue, New York, NY 10016, United States of America

British Library Cataloguing in Publication Data

Data available

Library of Congress Control Number: 2019936255

ISBN 978-0-19-881394-1

Printed and bound by
CPI Group (UK) Ltd, Croydon, CR0 4YY

This book is dedicated, as always, to my loving family:
my dear wife Sandy, and to Mike, Erica, and Nora.
They enrich my life and my writing.

Contents

Preface

'Physics' is a huge topic spanning big stretches of time, space, and ideas. The science itself is as old as the natural philosophy of the ancient Greeks, and as new as the most recent data from the National Aeronautics and Space Administration (NASA) or European Organization for Nuclear Research (CERN) and the latest technology. It is also a global undertaking. Many of its achievements such as the discovery of the Higgs boson and the detection of gravitational waves have been accomplished by large international teams of researchers.

What physics covers is even vaster than the Earth. One goal of physics is to understand the entire universe, from its origins to its ultimate fate, and from quarks to all spacetime. Through its sub-areas and partner areas such as elementary particles, condensed matter, and astrophysics, physics contributes to our understanding of nature at all its scales; to the technology that defines how we live; and to exploring the nature of life and living things including humanity itself.

My aim in this book is to help you the reader gain insight into this big subject. I hope to show you enough physics, and enough about physics and its history, for you to appreciate what physics covers; to grasp how physicists carry out research and why that research is important; to understand how and why society supports physics, and how physics affects society; and to contemplate how physics answers some of humanity's greatest questions, from 'How did it all begin?' to 'How can we sustain ourselves on Earth?'.

This overview draws on my career as a physics researcher, educator, and administrator, and on my extensive work in science outreach and popular science writing. My research experience includes industrial, academic, and government laboratories; 'small' physics with table-top equipment in my own lab, and 'big' physics at the Los Alamos National Laboratory and other major installations. I've visited places where physics history was made, such as the Mt Wilson observatory, where the expanding universe was discovered; the site of the first atomic bomb test at Alamogordo, New Mexico; and the Paris Observatory, where the speed of light was first measured.

In sharing my varied experiences and knowledge of physics, I've focused on making the subject accessible. My writing assumes minimal prior knowledge of physics, and I discuss it in descriptive and conceptual terms rather than mathematical ones.

If after finishing *Physics: A Very Short Introduction* you understand better what physics is about, and how and why physics and physicists operate as they do, this book has been a success. And if after finishing it you want to know more, or perhaps even want to pursue a career in physics or any science, then the book has been a double success.

Most of all, I simply hope that you enjoy the book.

Sidney Perkowitz
Atlanta, Georgia, and Seattle, Washington
2017–18

Acknowledgements

It is a pleasure to thank my commissioning editor at Oxford University Press, Latha Menon, who responded positively to my idea of a VSI on physics, shepherded it through the proposal process, and improved the manuscript through her editing; Jenny Nugee and Carrie Hickman, who efficiently responded to my queries as I wrote the book and who tracked down illustrations, respectively; and Joy Mellor for her effective copyediting. I also thank the anonymous reviewers for their helpful comments.

Special thanks go to my volunteer readers, Marc Merlin and Winston King, respectively Executive Director and Co-Director of the Atlanta Science Tavern. They read my manuscript at various stages, then used their deep knowledge of physics and of science outreach and teaching to suggest changes that made the book more lucid and more complete.

For their help in providing and explaining statistics about US federal support of physics, employment of physicists, and production of physics PhDs, I thank two staff members at the American Institute of Physics: Patrick Mulvey, Senior Survey Scientist, Statistical Research Center; and William Thomas, Science Policy Analyst, Government Relations Division.

Any errors in the book, however, are mine alone.

List of illustrations

Physics

List of abbreviations

a	acceleration
Al_2O_3	aluminum oxide
APS	American Physical Society
c	the speed of light
CAT	computerized axial tomography (scan)
CdSe	cadmium selenide
CERN	European Organization for Nuclear Research
CMB	cosmic microwave background
CO_2	carbon dioxide
DoD	US Department of Defense
DoE	US Department of Energy
E	energy
ESA	European Space Agency
F	force
fMRI	functional magnetic resonance imaging
GaN	gallium nitride
GPS	global positioning system
H	horizontal
ICBM	intercontinental ballistic missile
ISS	International Space Station
ITER	International Thermonuclear Experimental Reactor
LED	light emitting diode
LHC	Large Hadron Collider
LIGO	Laser Interferometer Gravitational-Wave Observatory
LQG	loop quantum gravity
m	mass
MIT	Massachusetts Institute of Technology

MRI	magnetic resonance imaging
NAS	National Academy of Sciences
NASA	National Aeronautics and Space Administration
NIF	National Ignition Facility
NMR	nuclear magnetic resonance
NRC	National Research Council
NSF	National Science Foundation
PET	positron-emission tomography
QED	quantum electrodynamics
qubit	quantum bit
QUEST	Quantum Entanglement Space Test
Rad Lab	MIT Radiation Laboratory
ROMY	Rotational Motions in Seismology
sonar	sound navigation and ranging
V	vertical
WIMP	weakly interacting matter particle

Physics

Chapter 1
It all began with the Greeks

Even if you're not a physicist and have never taken a physics course, physics is part of your life. Your smartphone uses semiconductor chips based on quantum mechanics; if you have ever had an X-ray or an MRI scan, you've benefited from medical techniques developed in physics labs; if you care about clean energy and maintaining the environment, or worry about the chances of nuclear war, physics hovers over those issues as well; and if you have ever looked at the stars in the night sky, and wondered as we always have how they and all creation arose, the science of physics has answers.

From consumer devices to research at the edge of the unknown, physics is woven into our daily activities, our civilization, and our highest aspirations. It is a foundational science that is massively supported by society and underpins other science and technology. Yet this essential human enterprise began long ago and at a small scale, in the minds of a handful of Greek thinkers who wanted to understand the world around them.

Curiosity and understanding

If there is a constant theme that runs through the history of physics, from its early days as natural philosophy practised by the ancient Greeks to today's intricate theories and apparatus, it has two strands: curiosity about the natural world, and the belief that we humans can comprehend it. Our early ancestors were surely astonished and awed by the natural phenomena around them, from the rising and setting of the sun to the ceaseless sound and motion of the ocean. Those feelings must have been accompanied by the desire to understand what they saw.

We still feel wonder at the beauty and workings of the universe around us, but with a difference. Now we can explain much, though not all, of what we take in with our senses or through specialized instruments. We continue to seek broader and deeper understanding, and today we have the tools to satisfy that curiosity, tools whose development is a large part of the history of physics and of all science.

Physics and nature

The connection between physics and nature is embedded in the very beginnings of the science, whose name comes from the Greek root *physis* meaning 'nature'; but the tools that formed modern physics took a long time to evolve. Our early ancestors believed the workings of nature to be under the arbitrary, sometimes capricious, control of gods such as Zeus, the chief Greek Olympian, who punished other gods and mortals when they challenged his authority. That belief in the gods came to be replaced with a set of concepts and techniques that would coalesce into physics, an integrated approach to exploring the universe.

Chief among these ideas was and still is the belief that we can grasp the workings of the world through the rational exercise of the human intellect. This was a major change from believing that the gods could do as they wished beyond human knowledge or control. Closely related is the idea of cause and effect, the conviction that a physical action in the world such as the application of a force gives a predictable outcome; and that the same cause always produces the same effect under equal conditions.

2

Other necessities to develop a science of physics were careful observation and record-keeping of natural events such as the regular motion of planets and stars. Later came quantitative analysis. Mathematics became the language of physics, the most powerful, concise, and exact way to manipulate data, express physical ideas, model the world, and predict its behaviour. Two other prerequisites were the notion of the experiment, an artificially limited portion of reality designed to test a particular idea, typically combining observation with the recording of quantitative data; and the development of a theoretical physics that uses mathematics to analyse and explain physical behaviour and experimental results.

These elements did not come quickly or in one place, but over millennia in different nations and cultures. In one example that became woven into the fabric of physics, before the Greeks the early Sumerians, Babylonians, and Egyptians developed measurement methods and applied mathematics. The Sumerians and Babylonians produced catalogues of stars, and the Egyptians cultivated practical mathematics to keep track of land allocations when the Nile flooded and erased boundary markings.

Other ancient civilizations around the world also developed aspects of physical and quantitative science, and technology that showed command of physical principles. The Chinese, for example, invented the abacus and the magnetic compass among other devices and processes; in South Asia, Indian mathematicians developed or widely transmitted fundamental mathematical ideas such as the concepts of zero as a number, and negative numbers; in Central America, the Mayan people developed sophisticated astronomical systems and calendars; and in South America, the Incas engineered an extensive road system along with aqueducts for water management.

However, the central idea that nature has 'laws' that operate without the intervention of the gods and could be rationally understood goes back to the early Greek philosophers, especially Aristotle who presented a world system in his work *Physics*. That title conveyed a different meaning than it does now. Natural philosophers like Aristotle were not like today's physicists. They did not necessarily carry out experiments or use mathematics, but they did seek laws of nature such as the laws of motion and of change, a major theme of *Physics*.

Moving objects

It seems inevitable that the study of motion was important to the Greek thinkers. Little on Earth or in the heavens is fixed and we live surrounded by change, as the Greek philosopher Heraclitus recognized when he wrote 'No man ever steps in the same river twice'. Also, we constantly see and viscerally experience the causes and effects of motion in daily life, such as the force needed to hurl a rock, and the differences in its path after throwing it with greater or lesser effort.

Aristotle's analysis of motion contained some insights, but his law of falling bodies showed the weakness in Greek thinking when it did not use empirical results to understand nature. His assertion that heavier bodies fall faster seems commonsensical but is incorrect. Only after much experimentation do we know that all material bodies fall at the same acceleration in a given gravitational field. Yet Greek thought yielded important ideas such as atomic theory, championed by Democritus of Abdera who summed it up in a single concise statement: 'Nothing exists except atoms and empty space, everything else is opinion'. And some early thinkers did carry out experiments and observations. For example, Archimedes discovered the principle of buoyancy in a liquid, and in the 2nd century CE, Claudius Ptolemy gathered data about the refraction or bending of light as it moves between different media.

Each succeeding century brought its own contributions to physics. Early work followed the writings of the Greek philosophers, especially Aristotle, but not all cultures extended these ideas. The Roman philosopher and poet Lucretius exerted great influence simply by explaining Greek science to his fellow citizens. His poem *On the Nature of Things* is famed for its clear presentation of Greek ideas such as atomic theory. The poem influenced Isaac Newton's understanding of falling bodies, and the Scottish mathematical biologist D'Arcy Thompson, author of the classic work *On Growth and Form* (1917) about the physical constraints on living things, saw Lucretius as providing something for every generation of scientist.

Greek and Roman contributions to physics are part of western culture, but beginning in the early 9th century CE, Middle Eastern culture took the lead in studying and transmitting physical ideas. Scholars in Baghdad translated the works of Aristotle and other Greeks into Arabic, and the Arab astronomer and mathematician Ibn al-Haytham (also known by the anglicized name, Alhazen) made strides in optics. Ibn al-Haytham showed that Ptolemy's analysis of refraction was wrong and rejected his idea that vision comes as light rays leave the human eye, in favour of the correct idea that vision arises as light rays enter the eye.

Further translations of the Arab and original Greek works initiated progress in physics in Europe during the Dark Ages. This slowly developed until the 16th and 17th centuries as natural philosophers moved away from Aristotelian physics and towards the quantitative analysis and experimentation that would come to underlie physics.

The Renaissance and the planets

Much of this progress occurred during the Renaissance, the period from the 14th to the 17th centuries when new cultural, artistic, and scientific ideas flowered in Europe. While some scholars have questioned the validity of isolating this period as special and its causes are debated, the times did throw up important figures across Europe who advanced physics and its methods.

One such, the Polish mathematician Nicolaus Copernicus, created a breakthrough in astronomy and cosmology, the Copernican Revolution. Ptolemy had earlier used astronomical data to build a model of the universe with the Earth as its centre, a geocentric view that was accepted for centuries. But in 1543, Copernicus' book *On the Revolutions of the Celestial Spheres* put the sun at the centre. With this heliocentric theory, Copernicus placed the observed planets into their correct order of distance from the sun, but he clung to Plato's belief that celestial orbits must be perfect circles. As with Ptolemy, that made it impossible to explain the observed planetary movements without unwieldy and arbitrary adjustments to their orbits that had no physical basis.

Fixing this problem fell to the German astronomer Johannes Kepler, who found that no amount of tinkering with a circular or oval orbit could reproduce the observed behaviour of Mars, but an elliptical orbit could. Kepler expressed his results in three laws of planetary motion: their orbits are elliptical, with the sun offset from the centre to one focus of the ellipse; the imaginary line joining a planet to the sun sweeps out equal areas in equal times; and the orbital period for each planet increases with its distance from the sun according to a specific relation. For instance, at 5.2 times further from the sun than the Earth, Jupiter takes 11.9 years to complete one orbit.

Later the great English scientist Isaac Newton would show that these apparently separate laws arise from one source, his law of universal gravitation. But before Newton could establish that and his laws of motion, the Italian scientist Galileo Galilei had to explore the basis of mechanics and of experimentation itself. Galileo was born in 1564, the year Michelangelo died, and died in 1642, the year Newton was born—an apt span for a polymath with artistic and musical ability whose approach to research shaped physics.

Originally intended for a medical career, Galileo instead became professor of mathematics at the University of Padua where, starting in 1610, he made and used astronomical telescopes. He discovered the four biggest moons of Jupiter, and observed that Venus displays phases like our moon, strong evidence for a heliocentric rather than a geocentric solar system. In 1632, his defence of the Copernican system in his *Dialogue Concerning the Two Chief World Systems* was seen by religious authorities as contrary to scripture. He was brought to trial and punished but still carried out ingenious experiments that overthrew Aristotle's ideas about falling bodies and established empirical results as essential for physics.

Gravitation, light, and Newton

Other pioneering thinkers such as the 17th-century French philosopher René Descartes continued to develop ideas about motion. His work influenced Isaac Newton, the towering 17th-century figure considered one of the greatest scientists of all time. As a child, Newton constructed mechanical devices such as a water clock, then studied science and mathematics at Cambridge University. With this background, his scientific career displayed his great skills in both experiment and theory.

In 1687, Newton's seminal book *Mathematical Principles of Natural Philosophy* (usually called simply the *Principia* from its original Latin title *Philosophiæ Naturalis Principia Mathematica*) put forth his three laws of motion, which explained all dynamic phenomena until relativity and quantum mechanics appeared; and his law of gravitation, which shows how to calculate the gravitational attraction between any two bodies (Figure 1). This result embodied Kepler's three laws and made it possible to

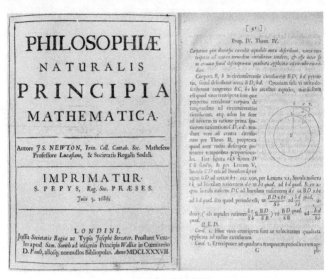

1. **The cover and a page from Isaac Newton's *Principia* (1687).**

forecast the behaviour of celestial as well as earthly bodies. The law is simple enough for beginning physics students, yet it nearly perfectly described celestial motion until Albert Einstein developed general relativity in 1915 to give more accurate predictions.

Newton also tackled optics and the nature of light. In 1669, he invented an improved 'reflector' type of telescope that used a curved mirror instead of lenses. His design is still preferred for serious astronomy, exemplified in giant ground-based units and NASA's Hubble space telescope.

To study light itself, in 1666 Newton carried out a famous, seemingly simple experiment; he inserted a glass prism purchased at the Sturbridge Fair into a ray of sunlight to produce a rainbow, splitting what had been thought to be pure white light into component coloured rays. These travelled in divergent paths because of refraction, which Newton called 'refrangibility'. As he explained,

> To the same degree of refrangibility ever belongs the same
> colour, and to the same colour ever belongs the same degree of
> refrangibility. The least refrangible rays are all disposed to exhibit a
> red colour…the most refrangible rays are all disposed to exhibit
> a deep violet colour…

To confirm this insight, Newton carried out another experiment
that recombined the coloured rays into white light.

Newton's work with light was the latest attempt to explain
this intangible element of nature, whose essential constitution
and speed had long been mysteries. The ancient Greeks
thought it travelled at infinite speed, but in 1676 the Danish
astronomer Ole Roemer calculated a finite value from
astronomical data that came within 20 per cent of the correct
value, 300,000 kilometres/second (186,000 miles/second). Light
however displayed a seeming paradox. It would bend around
obstacles like ocean waves bending around a jetty, suggesting that
it is wave-like; but its rays travelled in straight lines, suggesting
that it is particulate. The latter was a main reason that in his book
Opticks (1704) Newton described light as made of 'corpuscles'
that are refracted according to mechanical principles.

Newton's theory was not the final word on light; nevertheless, with
his combination of crucial experiments and rigorous mathematics,
as the science historian J. L. Heilbron observes, he 'completed one
full turn of the helix of scientific advance…Newton was the
Napoleon of the Scientific Revolution' who after Galileo set the
standard for the modern approach in physics.

Electric and magnetic fluids

Along with light, natural philosophers explored electricity and
magnetism as other weightless or 'imponderable' parts of nature.
The early Greek philosopher Thales of Miletus apparently knew of
the magnetic properties of lodestone and the static electrical

properties of amber when rubbed (the electron, with its negative electrical charge, is named after the Greek word for amber). Centuries later, the English physician William Gilbert and the English astronomer Edmond Halley of comet fame studied lodestones and the Earth's magnetism, and Newton himself displayed static electricity to the Royal Society.

The 18th century saw great advances in electricity: the use of the Leyden jar to store electrical charge for new experiments (in one amusing demonstration, the experimenter passed a spark through a long line of monks holding hands, making them simultaneously jump); the association of lightning with electricity by the American statesman and scientist Benjamin Franklin, and his proposal that electricity is a fluid (other theories postulated two types of electrical fluid); Alessandro Volta's invention of the 'voltaic pile' or electric battery in 1800; and the surprising discovery that electricity and life are linked, when in 1780 the Italian researcher Luigi Galvani found that electricity made a dead frog's leg twitch (Mary Shelley named 'galvanism' as a mechanism to animate dead matter in the 1831 edition of her story *Frankenstein*).

The age of correlation

By the end of the 18th century, great progress had been made in understanding the physical world. Newton's mechanics described motion; his corpuscles explained the behaviour of light; electricity and magnetism had been explored, and, it was theorized, were carried by weightless fluids; and thermal effects were thought to arise from caloric, another weightless fluid that flowed from hot to cold bodies. But many of these ideas were upset by new data and theories in the 19th century.

That century has been called the 'age of correlation' where ideas came together to form what is now called classical physics, a unified approach to analysing physical phenomena. Perhaps this coherence also helped establish physics as a well-defined science. Its practitioners were for the first time recognized as 'physicists', a title proposed in 1840, and physics curricula were installed in schools and universities.

Classical physics would include new correlations among electricity, magnetism, and light, but light itself had to be better understood. Though Newton's corpuscular theory was widely accepted, in 1690 the Dutch scientist Christiaan Huygens proposed instead that light consists of waves travelling in a space-filling 'ether'. Then in the early 1800s, the English polymath and physician Thomas Young convincingly showed that light is a wave. He passed light through two holes in a screen, producing a pattern of alternating illuminated and dark areas that could only arise from constructive and destructive interference between light waves. Later results showed how interference between waves could produce straight light rays, overturning Newton's main objection to the wave theory.

However, the nature of these light waves remained unknown. A clue came in 1831 when the outstanding English experimentalist Michael Faraday showed that a changing magnetic field could induce an electric current, giving a new connection between electricity and magnetism. Then in 1865 the Scottish mathematical physicist James Clerk Maxwell merged Faraday's result with everything else known about electricity and magnetism into a single entity, electromagnetism. Later reduced to four mathematical expressions known as Maxwell's Equations, this theory unexpectedly showed that electromagnetic waves could be generated and would travel at the known speed of light. Experiments confirmed that light is indeed an electromagnetic wave, at last establishing its true nature.

These results eliminated notions of electric and magnetic fluids, and 19th-century physicists abolished another imponderable when they asked 'what is heat?' In 1798, the American-born scientist Benjamin Thompson, Count Rumford, observed that the friction produced by grinding cannon barrels generated unlimited amounts of heat. He concluded that heat 'cannot possibly be a material substance'—which eliminated caloric—but must be related to motion. His surmise was confirmed when measurements established that a given amount of work always created the same amount of heat, and provided a numerical value for this mechanical equivalent of heat.

Further development of thermodynamics—the theory of heat, energy, and work that is basic to physics—began with the French engineer Sadi Carnot. Interested in the performance of steam engines, in 1824 he derived a general rule for the maximum efficiency of any engine driven by heat. His analysis led to a new physical quantity, entropy, which measures how much thermal energy is unavailable to perform work in any thermodynamic process or engine that uses heat. Later efforts by other researchers produced the powerful Three Laws of Thermodynamics, including the conservation of energy, and the tendency of entropy always to increase. This points to the eventual decline of any system, including the entire universe, and also provides an 'arrow of time', a one-way indicator of the direction in which time flows.

Conservation of energy has become a bedrock idea of physics. The recognition that heat is related to motion and then specifically to molecular motion was also hugely important because it linked Newtonian mechanics to thermal behaviour. When gas molecules were analysed as a swarm of tiny moving masses, it became clear that the temperature of a system directly reflects the kinetic energy of its molecules. This kinetic theory of gases was pioneered by the Austrian physicist Ludwig Boltzmann and provided a new approach to the atomic-scale world—a world that would receive increasing scrutiny in the 20th century.

With its new general principles and theories, 19th-century physics overthrew all the old imponderables except the ether. If the ether were to support electromagnetic waves travelling at high speed, and yet at the same time not resist the motion of the planets, it would have to possess completely contradictory properties and so its nature remained a puzzle.

Otherwise, classical physics could claim significant successes for its explanatory power. It also influenced the technology of the time: after James Watt improved the steam engine in 1781, applied thermal physics further enhanced the operation of this power source for the Industrial Revolution; Count Rumford used his knowledge of heat to invent the kitchen stove, elevating cooking from the use of an open fire to a more refined art; and Faraday's work laid the basis for electric motors and generators.

20th-century surprises

Given these achievements, some 19th-century physicists felt that physics was essentially complete. In 1896, Albert Michelson, who in 1907 would become America's first Nobel Laureate in science, wrote:

> While it is never safe to affirm that the future of Physical Science has no marvels in store…it seems probable that most of the grand underlying principles have been firmly established…further advances are to be sought chiefly in the rigorous application of these principles to all the phenomena which come under our notice.

Ironically, Michelson's own precise measurements of the speed of light, and other research as the 19th century became the 20th, discovered unexpected physics of the microscopic and macroscopic worlds. These findings changed classical physics into what is still called 'modern' physics, though it goes back to research from 1900 to the 1930s.

Some of the new results came from a device invented by the English physicist William Crookes in the late 19th century, a partially evacuated glass tube with metal electrodes. With a high voltage applied to the electrodes, an electrical discharge then called 'cathode rays' streamed between them. In 1895, the German physicist Wilhelm Röntgen found that a Crookes tube also produced 'X-rays', an unknown type of invisible radiation that penetrated solid matter, later determined to be short wavelength electromagnetic waves.

Two years later, the English physicist J. J. Thomson found that cathode rays were really streams of negatively charged particles later called electrons, the first elementary particle to be discovered. Then in 1911 the New Zealand physicist Ernest Rutherford showed that an atom consists of electrons surrounding a tiny nucleus, itself eventually found to be made of positively charged protons and electrically neutral neutrons.

In other contemporary work, in 1896 the French physicist Henri Becquerel, the Polish-French researcher Marie Sklodowska Curie, and her husband Pierre Curie began exploring radioactivity, where certain elements spontaneously emit radiation, changing as they do—radium for instance turns into lead. Later these so-called alpha, beta, and gamma emissions were shown to consist respectively of protons and neutrons combined into helium nuclei, electrons, and electromagnetic radiation at wavelengths shorter than those of X-rays.

The newly discovered small world of electrons, nuclei, and atoms showed an unexpected aspect; it is quantized, meaning it operates with energies that come only in specific discrete amounts, not in a continuous flow. This possibility was first put forward in 1900 by the German physicist Max Planck to resolve anomalies in the behaviour of light from hot glowing sources. The idea seemed outlandish and Planck himself initially viewed it sceptically, but it proved valid and indeed essential to describe light and matter.

Light is quantized as separate small packets of energy called photons, with the energy proportional to the frequency of the light, as proposed by Einstein in 1905 to explain how light ejects electrons from certain metals. In matter, the energy levels of electrons in atoms are also quantized, as the Danish physicist Niels Bohr proposed in 1913 for hydrogen. The introduction of the photon re-opened the question whether light is a particle or a wave, and quantum mechanics has other apparent paradoxes; yet together with its extended form, quantum field theory, it has been eminently successful with important technological outcomes.

While physicists absorbed the implications of the quantum, other results changed our view of the ether and the universe. In 1887, Michelson with Edward Morley carried out a sensitive experiment that sought small changes in the speed of light to determine whether the Earth moves through the ether, and found no evidence of such motion. This troubling result was explained eighteen years later by Einstein's special relativity. In making space and time variable and in taking the speed of light as having the same value relative to any moving observer, the theory eliminated any need for the ether, the last 19th-century imponderable.

In 1915, Einstein extended his theorizing to create general relativity, which describes gravity as resulting when a mass distorts spacetime, the four-dimensional fabric of the universe, rather than as a force acting over distance as Newton had surmised. Einstein's result was soon experimentally confirmed through its correct predictions of the behaviour of light and of planets under gravity.

Besides these theoretical tools, advanced experimental methods provided new insights. In 1911, the Dutch researcher Heike Kamerlingh Onnes discovered that some metals become 'superconducting' and lose all electrical resistance when cooled

2. The 2.5-metre (100-inch) reflector telescope installed at Mt Wilson, CA, in 1917.

near absolute zero. In 1917 the world's then largest reflecting telescope was installed at Mt Wilson, California, enabling the American astronomer Edwin Hubble to begin establishing the size of the universe (Figure 2). He found that the Andromeda Galaxy lay a million light years away (later corrected to two million light years. A light year is the distance light travels in a year, about 10^{13} kilometres). Equally significant for our understanding of the universe, his data showed that it is expanding—a result that the Belgian priest and astronomer Georges Lemaître had considered earlier.

At the other end of the scale, in 1932 the American physicists Ernest Lawrence and M. Stanley Livingston invented the cyclotron. This device accelerated sub-atomic particles around a circular track until they reached high energies and could probe atomic nuclei or collide and create new particles. And, in 1938, another probe of small-scale nature changed physics, the course of World War II, and the post-war world. After the German chemists Otto Hahn and Fritz Strassman bombarded uranium with neutrons, the Austrian-German physicist Lise Meitner and her nephew Otto Frisch confirmed that this had made atomic nuclei split or fission into smaller pieces. Nuclear fission, with its enormous energy release, was soon exploited in the Manhattan Project to make the two atomic bombs the United States dropped on Japan in 1945 that ended World War II.

The physicist's war

That war set new directions for physics and its role in society. World War I had been 'the chemist's war' because of its use of poison gas and high explosives; World War II was 'the physicist's war', with its use of weapons and technology with physics roots such as the atomic bomb, radar, and the Nazi V-2 rocket. After the war ended in 1945, these provided starting points for nuclear power, space flight, the laser, and other scientific and civilian spinoffs. Relatively unharmed by the war compared to other nations, the US government and economy had the means to strongly invest in physics and its applications.

Physics was supported in other countries as well for reasons of national defence on both sides of the Cold War, the period of hostility short of open conflict between the US and the Soviet Union and their allies from 1945 to 1991. To some observers, this produced an undesirable bias towards military and nationalistic applications, and a need for scientific secrecy that harmed physics; to others, physics broadly benefited from government funding when it was coupled with academic research and industrial support.

With the Cold War now gone, and growing global economic and industrial strength, physics is returning to its earlier international roots as it thrives throughout Europe and Asia as well as the US. Besides these separate efforts, major collaborations have developed among nations, such as the particle physics research carried out at the enormous Large Hadron Collider (LHC) near Geneva, Switzerland, a particle accelerator operated by the multi-nation CERN (Figure 3); or the International Space Station (ISS), a large US–Russian artificial satellite in low Earth orbit that supports research in space science and technology for a variety of nations.

The next correlation

Now in the 21st century, results from machines like the LHC and progress in theory are providing new impetus to seek another correlation for fundamental physics. Decades of particle physics experiments and theory have yielded the Standard Model, based on quantum field theory. It organizes all the known elementary particles into classes and explains the origin of three of the four fundamental forces that make the universe work, from quarks to galaxies—the electromagnetic interaction, and the so-called weak and strong interactions in atomic nuclei. But it excludes gravity, which is separately described by general relativity.

The next great hoped-for correlation would quantize gravity and fit it into the scheme with the other interactions, leading to an all-encompassing Theory of Everything that describes reality at every scale. The leading candidate has been string theory, in which tiny one-dimensional objects called strings replace point-like elementary particles. But physicists do not see how this theory can be experimentally tested because that would require a particle accelerator so much bigger than the LHC that it seems impossible to build. Still, faced with the absence of any other definite

3. Aerial view of the Large Hadron Collider at the Franco-Swiss border.

theoretical successes, some physicists suggest we should bypass experimental confirmation and simply accept string theory and the related idea that reality consists of multiple universes, the 'multiverse'—a radical change in how physics has operated for centuries.

Another possibility is to more vigorously pursue a rival approach, loop quantum gravity (LQG), which string theory has overshadowed. In LQG, space and time are themselves quantized into discrete units rather than forming smooth continuous media as we now think of them. Recent research is showing some connections between LQG and string theory. Perhaps this will lead to results that would not require discarding the core idea of experimental validation.

In a future history of physics that looks back on our time, this may be the century where the search for a unifying Theory of Everything either preserved and extended the established way of doing physics or took physics in new directions.

Chapter 2
What physics covers
and what it doesn't

Look up the word 'physics' online or in dictionaries and you'll likely see a definition more or less like this: 'Physics is the study of matter, energy, and the interaction between them'. This is concise and seemingly clear, but not really, because 'matter' and 'energy' are harder to define than you might think, so that specifying them with precision helps to define physics itself. Besides, physics is a sprawling and dynamic science, and this basic definition needs to be amplified to truly reflect what physics is all about.

Matter and energy

Nevertheless, the definition has a big plus: it is completely general, because from the viewpoint of physics, matter and energy, existing within space and time, represent the entirety of the universe. It is worth stressing that the definition therefore shows that physics treats reality as purely materialistic. Unlike entities such as 'mind' and 'soul', matter and energy can be directly sensed or detected and objectively measured by observers. Some philosophers have held instead a dualistic view, the idea that the universe contains both physical and non-physical things. Plato believed the soul could migrate from body to body; Descartes held that matter with its spatial extent differs from mind, which is non-spatial and has the ability to think. Physics as practised explicitly does not follow any such dualism.

But definitions that refer only to matter and energy tell us little about the texture of physics and how it works. Instead, physics can be defined and described by scale, that is, how it deals with different aspects of the universe by their size; by its sub-areas such as mechanics and nuclear physics; by the research methods it uses; by intent, meaning is it 'pure' or 'applied'; and finally by how it is actually practised, which illuminates its development over time. Still all these depend on matter and energy as central concepts.

About matter

The classical definition of matter is that it occupies space and has mass. These may seem self-explanatory properties but space and mass have their own complexities. As a basis for his theories of mechanics and gravity, in his *Principia* Newton defined an 'absolute space' that 'in its own nature, without regard to anything external, remains always similar and immovable' and an 'absolute, true and mathematical time' that 'from its own nature flows equably without regard to anything external'. Einstein's relativity upset this view by showing that space and time are not separate things but are connected to form 'spacetime', a unified four-dimensional entity. Moreover, his theory predicts and experiment confirms that spacetime is not fixed or absolute, but changes through 'length contraction' and 'time dilation', effects seen by observers moving at different speeds.

Special relativity also complicates the notion of mass itself. It predicts (and experiment confirms) that the mass of a moving body as seen by a fixed observer increases with the speed of the body. Only the body's so-called rest mass, measured by an observer located on the body, remains unchanged. Further, even for a body moving too slowly to bring in relativistic effects, there are different ways to define its mass. The inertial mass of a body represents its resistance to acceleration by a force, the reason it is harder to push a big stalled automobile than a small one; the gravitational mass of a body determines the strength of its gravitational interaction with other bodies (measurements show that the two masses are equal—the reason that all objects fall at the same rate and an important result for general relativity).

Our view of matter has also changed. Its classic definition characterizes three different macroscopic states: solids, which hold a fixed shape and volume unless deliberately deformed; liquids, which take the shape of their container; and gases, which fill a container. All three exist on our planet and throughout the universe.

A fourth state was called 'radiant matter' in 1879 when William Crookes first observed it in the tube he had invented. He was seeing a plasma, a hot gas whose atoms had been ionized into separate positive nuclei and negative electrons (in this case, air heated by high voltage). Hot plasmas are not prevalent on Earth except when generated by lightning, but we make them artificially in neon signs and plasma televisions. They are, however, widespread throughout the universe inside active stars such as our sun, which are made of extremely hot plasma.

By the early 20th century, physics could begin to explain all four states of matter in terms of atoms made of protons, neutrons, and electrons, and how the atoms are arranged. Now we have found quarks, the constituents of protons and neutrons, and other elementary particles including the Higgs boson which gives some elementary particles their mass, bringing us closer to understanding the underlying nature of matter. We have also found additional states of matter such as Bose–Einstein condensates, where a group of extremely cold atoms enters into a shared quantum state; the super-dense cores of certain dead stars made of neutrons; and dark matter, the invisible entity that makes up much of the universe but whose nature remains unknown.

About energy

Like matter, energy has multiple meanings; in physics, it is the ability to do work, that is, move a force—a push or a pull—through a distance. Think of an accident where automobile A collides with automobile B. That pushes or exerts a force against the metal of auto B, crumpling it as the force moves through a distance to perform work on B. This work came from the kinetic energy or energy of motion that auto A carried. Any moving object has kinetic energy—a thrown baseball, a molecule in a gas, or an electron travelling through a wire. The idea can also be extended to include motion that does not involve objects with mass. 'Radiant energy' is carried by light and other electromagnetic waves. These exert forces on charged particles such as electrons that make them move, thereby performing work.

A broader definition brings in the concept of potential energy, which is work that has been stored in a system by changing its configuration. For example, it takes work to raise a boulder from ground level to a cliff top. That work can be retrieved as kinetic energy if the boulder is tipped over the edge to fall, then changed again to work if the mass smashes through some structure below. Potential energy also exists in the chemical bonds between atoms, and in atomic nuclei where it is released if the mass (m) is converted into a large amount of energy (E) according to Einstein's equation $E = mc^2$ where c is the speed of light. Another form of energy is dark energy, an entity discovered in 1998 that seems to fill all space and speeds up the expansion of the universe.

Size matters

One goal of physics is the reductionist one of finding the ultimate building blocks of reality that comprise the universe up to its biggest structures. This represents a staggering range of sizes. Protons, neutrons, electrons, and other fundamental particles are so tiny that they can be treated as mathematical points. Atoms are bigger but still minute, typically less than a nanometre across (a nanometre is a billionth of a metre, 10^{-9} metre). At the far end of the scale, huge cosmic entities are measured in light years. Our own Milky Way galaxy is 100,000 light years across.

Quarks to atoms to galaxies, the components of the universe span many orders of magnitude in size (Figure 4). Physicists dream of a Theory of Everything that explains reality at all scales; but since that does not exist, different approaches have evolved to deal with this enormous range in size, in three parts with overlapping boundaries: the small or microscopic to sub-microscopic—elementary particles, atoms, and molecules; the mid-size or mesoscopic, from the human to the planetary scale; and the large or macroscopic, consisting of stars, galaxies, the entire universe, and its spacetime.

Early natural philosophers could observe and up to a point manipulate mid-size physical phenomena, and though they could not reach celestial bodies, they could observe them as well at the mid-size to large scales. Without proper instruments, however, they could not examine sub-microscopic things. Later centuries brought advanced technology but this too worked mostly at the mid-size to large levels as telescopes enhanced astronomical observation, Newton examined light with a glass prism, Faraday explored electromagnetism with batteries and wires, and Carnot investigated heat by considering the steam engine.

These efforts led to a classical physics that explained everyday phenomena and those of bigger scale on and in the Earth, its oceans, and atmosphere, along with celestial activity. Newton's mechanics and gravitation accounted for ocean tides as well as planetary orbits; work by the Irish physicist John Tyndall and later by the British physicist Lord Rayleigh showed that short wavelength blue rays in sunlight scatter preferentially in the

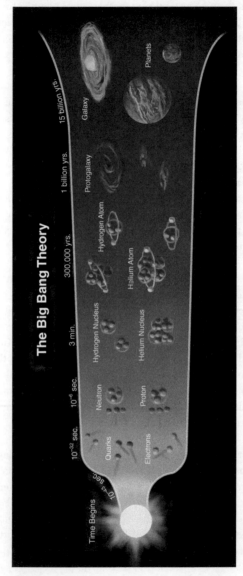

4. Evolution of the universe from quarks and electrons to its present size.

Earth's atmosphere, making the sky blue; and the British seismologist Richard Dixon Oldham analysed mechanical earthquake waves to examine our planet's internal structure.

Progress was slower at the smallest scale. Democritus had presented an atomic theory in 465 BCE in which matter consists of tiny solid indivisible particles in constant motion, but it took later breakthroughs to probe the atomic and sub-atomic world. The optical microscope, invented in the late 16th century, could not do so; it is limited by the wavelength of visible light to a resolution of 200 nanometres, far bigger than an atom. In 1912, however, the British father–son physicists William Lawrence Bragg and William Henry Bragg earned a Nobel Prize for demonstrating that X-rays interacting with the atoms in a crystal are deflected to produce a pattern that gives the atomic locations. X-ray scattering became invaluable to map atoms and molecules in inorganic and biological materials; in 1952 the British physical chemist Rosalind Franklin used X-rays to examine DNA, with results that led James Watson, Francis Crick, and Maurice Wilkins to establish its double helix structure.

Further progress came with the inventions of the electron microscope in the 1920s, which uses the wave-like quantum mechanical properties of electrons; and of the scanning tunnelling microscope in 1981, which also uses a quantum mechanical effect and earned its inventors a Nobel Prize. Today these and other techniques achieve resolutions of a fraction of a nanometre, comparable to the size of a hydrogen atom. These non-optical forms of microscopy routinely image and manipulate individual atoms, typically located on the surfaces of materials.

At the sub-atomic level, it took the development of particle accelerators that speed up electrons and protons until they have enough energy to break atomic nuclei into their components, or to collide and generate other elementary particles. Since the invention of a table-top cyclotron in 1932, these accelerators have grown hugely in scale and power, as shown by two machines where different quarks were discovered—SLAC at Stanford University, the Stanford Linear Accelerator 3.2 kilometres long, and the Tevatron at the Fermi National Accelerator Laboratory, a circular ring 6.9 kilometres around—culminating in the huge LHC where the Higgs boson was found in 2012 (see Figure 3).

Along with these experimental tools, quantum theory and its extension quantum field theory were developed between 1900 and the 1940s. The combination of this theory and experiment yielded the Standard Model. Quantum theory also successfully described atoms as individuals and when they are arranged into bulk matter.

The macroscopic world requires its own apparatus. Astronomy depended on the naked eye until Galileo examined the skies through a small telescope in 1610. Starting in 1917, with the Mt Wilson telescope that used a 2.5-metre mirror (see Figure 2), now we have telescopes with mirrors up to 10.4 metres across. In the 19th century, telescopes became analytical tools when spectrometers were added to break astronomical light into its component wavelengths. That was the birth of astrophysics, because the spectrum of a star or other astronomical body tells us about its atomic constitution and its motion.

With the advent of space travel, astrophysics and planetary science entered a new phase. Space technology put NASA's Hubble telescope, the Planck satellite from the European Space Agency (ESA), and other devices into orbit to gather images and spectra from gamma rays through visible light to microwaves, without interference from the Earth's atmosphere. We have also taken close-up views of celestial bodies and even directly sampled them. NASA put people on the moon in 1969 and returned them to Earth with moon rocks, and later sent planetary rovers to examine the soil of Mars for signs of water, considered essential to support life. ESA landed a probe on the surface of Saturn's moon Titan in 2005, and in 2014 put a robot lander on a comet for the first time.

Space science, and astronomical phenomena in and beyond our solar system, require the theory of relativity to describe objects moving at or near the speed of light and the gravitation that forms cosmic structures. General relativity also predicts the existence of black holes, regions of space where extremely dense matter creates gravity so powerful that nothing, neither matter nor light, can escape. Together with quantum physics, nuclear and elementary particle physics, general relativity is necessary to understand the development of the universe and its stars and to probe black holes, dark matter, and dark energy.

In studying the small and the large, physics has created a dynamic theoretical narrative for the transition of one to the other, which is the history of the universe (see Figure 4). Present theory does not quite go back to the birth of the universe in the Big Bang 13.8 billion years ago but begins a mere 10^{-43} seconds after that, when energy and elementary particles were first formed. The theory traces many details of how these evolved into the universe we see today, and projects whether and how the universe will end billions of years from now.

Looking at physics through the broad categories of scale shows the cosmic sweep of the science in space and time. The categories of scale also correlate with the development of sub-areas in physics, some going back centuries and some much newer, which guide researchers to study nature at a finer grain.

Physics subdivided

The Greek natural philosophers wished to know about all nature. Aristotle's view of 'physics' included biology as well, for instance. This gave a kind of unity to these early efforts, but as humanity gained more scientific knowledge towards the end of the Renaissance, the study of nature split in specialized directions leading to modern physics, chemistry, biology, and so on. Eventually the sciences themselves divided into sub-areas that depended on the scale being studied and the research methods used.

In physics, those natural phenomena that were accessible and could be experienced at the human level developed first—mechanics, heat, sound, light, electricity, and magnetism—constituting classical 19th-century physics. By now this is established science that remains important because it accurately describes most of the phenomena around us, such as ordinary objects moving slowly compared to the speed of light. Research based on classical physics continues to develop insights, for instance in chaos theory, which shows that systems that follow classical mechanics can produce unexpected behaviour depending on their starting conditions, and in applied physics.

Classical physics also remains embedded in physics education. It introduces students to ideas like matter, energy, and force, and helps them develop physical intuition before tackling the abstract mathematical theories of modern physics. A typical undergraduate physics curriculum includes classical mechanics, heat, and electricity and magnetism, then relativity, quantum mechanics, mathematical and computational methods, and laboratory techniques. This provides the background for students to choose future paths for employment or for graduate school, and to choose between theory and experiment, and between pure and applied physics.

The areas that developed after the rise of classical physics are the most active arenas for research and raise questions at the boundaries of known physics. At the small scale, these include nuclear and particle physics, the physics of condensed matter (solids and liquids), and studies of the still mysterious quantum properties of light. These require exotic experimental capabilities like particle accelerators, temperatures near absolute zero, and special lasers; and all depend on quantum physics to understand experimental results and form new ideas. At the large scale, the research areas are astronomy, astrophysics, and cosmology—among the oldest of human scientific interests but now radically transformed with new telescopes and our entry into space, and with relativity, quantum physics, and nuclear and particle physics as theoretical background.

Theory, experiment, and more

Within each physics sub-area, there are physicists who gather data and others who devise theories to explain experimental results. Theory and experiment are the essential twins that drive physics but neither arose without considerable development. Before there was experiment, there was observation of natural activities such as the motion of celestial objects. Recorded and compiled, these observations made a data set that could be analysed, as Ptolemy had done though his geocentric model was incorrect.

New instruments enhanced the scope and accuracy of observational data. Even before the telescope was invented, the 16th-century Danish astronomer Tycho Brahe accurately measured the angular positions of celestial objects with what were essentially big protractors. That data allowed Brahe's assistant Johannes Kepler to derive his three laws of planetary motion. Later, telescopes yielded better observations and new discoveries. The planet Uranus is barely visible to the naked eye but the German-British astronomer William Herschel saw it through a telescope in 1781; the existence of Neptune, also invisible by eye, was confirmed by telescope in 1846.

Observational science takes nature as it comes and records what is seen. Experiment goes further, isolating and manipulating natural phenomena so they can be definitively examined. To test his ideas about light, the Arab scientist Ibn al-Haytham (see also Chapters 1 and 3) hung two lanterns at different heights outside a darkened room with a hole in one wall. Uncovering and covering the lanterns, he saw a spot of light correspondingly appear and disappear on the wall of the room, and noted that the spot for each lantern lay on a straight path back through the hole to that lantern. This experiment went far towards confirming that vision is due to light emitted by sources, not by the human eye, and that light rays travel in straight lines.

In 1638, Galileo further extended the scientific revolution that Copernicus initiated when he showed the importance of experiment followed by mathematical analysis. His book *Discourses and Mathematical Demonstrations Relating to Two New Sciences* describes his refutation of Aristotle's assertion that heavier objects fall faster. Galileo did this by a clever indirect method, because even had he dropped objects from the leaning Tower of Pisa as legend has it, they would have fallen too fast to be timed by available clocks. Instead he timed bronze balls rolling down inclined ramps more slowly than in free fall. Galileo's analysis showed that the acceleration under gravity is the same regardless of the weight of the ball. This is also true for objects in free fall, he argued, because that is exactly like motion on a ramp inclined at 90 degrees.

Decades later, Isaac Newton amplified Galileo's approach. His *Principia* expressed his belief that natural philosophers should combine data with mathematics to explain nature. Drawing on earlier results from Galileo and Descartes about moving bodies, and using geometry and the powerful tool of calculus he had invented (also invented independently by the German mathematician and philosopher Gottfried Leibniz), he derived the key principles of mechanics. His three laws of motion (best known is $F = ma$, force = mass × acceleration) and law of universal gravitation represent the first sweeping theories that explain a large part of natural behaviour on Earth and in the heavens (see Figure 1).

After Newton, theoretical physics continued to develop into a specific field within physics, with 19th-century researchers like the Scottish physicist James Clerk Maxwell specializing in the area. In one example, he used mathematical analysis and physical insight to show that the rings of Saturn could not be solid but must be made of separate components held together by gravity, as observation later confirmed. His mathematical skills led to his greatest achievement, a unified theory for electromagnetism based on experimental results that gave an unexpected bonus in showing that light is an electromagnetic wave.

The complementary roles of theory and experiment were clearly displayed as the 20th century began. At the First International Congress of Physics held in Paris in 1900, with over 800 attendees, both theorists and experimentalists presented keynote addresses. In 1911, at the first of a series of conferences sponsored by the industrialist Ernest Solvay that invited leading physicists to discuss the weighty scientific questions of the day, there was a nearly equal balance between theoretical and experimental physicists (but not between genders; Madame Curie was the sole woman among twenty-four invitees).

The symbiotic relationship among observation, experiment, and theory continues today, but few contemporary physicists combine experiment and theory as did Galileo and Newton. The scope and complexity of modern laboratories and theories force a physicist to choose one path or the other, and today a physicist can also choose computational physics. Since their inception computers have been useful for physics (and all science) to record and process data and solve mathematical problems. In computational physics, these consist of equations that correctly describe a real physical system but are hard to solve by the usual mathematical methods—typically because the system is non-linear or has many components, as occurs in general relativity, fluid flow, and condensed matter physics.

In that case, computers carry out accurate solutions, which amounts to numerically modelling or simulating the system.

The simulation parameters can be varied to explore what the mathematics predicts, later to be confirmed by experiment; or the simulation may stand in for an actual close-up view that would be physically impossible. That occurred in 2015, when the massive Laser Interferometer Gravitational-Wave Observatory (LIGO) detectors on Earth sensed for the first time the existence of gravitational waves. This confirmed Einstein's century-old prediction from general relativity that gravitational effects are carried by ripples in spacetime at the speed of light. It took a computational simulation of black hole behaviour to confirm that the detected signal came from a collision between two black holes 1.4 billion light years distant.

Computational physics is considered a branch of theoretical physics, or a 'third way' besides theory and experiment. More controversially, some physicists speak of 'computer experiments' that can replace real-world experiments. In any case, this area has growing influence.

Another change in physics is the appearance of decentralized research. Though some big physics projects are carried out at centres like CERN, others are performed by a global network of collaborating physicists tied together through personal computers and the Internet, as was done with LIGO. Network tools also allow anyone, physicist or not, to advance research. The CERN 'volunteer computing' website provides, as it states, 'a wonderful opportunity for YOU to get directly involved in cutting edge scientific research' by allowing one's personal computer to help analyse data from the LHC, a significant resource when thousands contribute. NASA's 'citizen scientists' website offers the chance to examine images of the exotic material aerogel to find interstellar dust particles trapped in it during the NASA Stardust mission of 2004; and as part of the 'Zooniverse' project at Oxford University, the opportunity to categorize by shape the galaxies seen by the Hubble Telescope.

Pure vs applied

The Greek natural philosophers preferred theorizing about nature to exploring it with experiments, nor did they think much about applying their knowledge to create actual devices for people to use. That changed with the Industrial Revolution in the 18th and 19th centuries, during which practical machines and processes benefited from knowledge of physics.

Initially there were no industrial laboratories that applied physics principles to improve processes or create new products. However physics research soon led to results that changed society and engendered new industries, with such efforts as exploring the thermodynamics of evaporating gases to create cold temperatures, which led to refrigeration technology in the late 19th century; and the exploitation of electricity and electromagnetism by Samuel F. B. Morse to invent the telegraph in 1844, by Alexander Graham Bell to create the telephone in 1876, by Guglielmo Marconi to project radio waves across the Atlantic Ocean in 1901, and by various inventors whose work produced electronically transmitted television in 1926.

Now many enterprises rely on applied physics and the engineering it supports. Examples funded by governments include exploration of space by NASA, ESA, and other national programmes; nuclear weapons development in the US and elsewhere; and the satellite monitoring of our planet's land areas, oceans, and atmosphere.

On the commercial side, physics principles have led to such pivotal devices and systems as the computer chip, the laser, radar, and magnetic resonance imaging (MRI), which lie at the heart of high-tech areas like digital electronics, photonics technology that manipulates light, and medical and consumer technology. The areas of condensed matter and materials physics, which among other research study technologically important materials like semiconductors, play big roles in bridging the gap between basic physics and device engineering. Condensed matter physics was the largest sub-area among US physics PhDs awarded in 2013 and 2014, and, together with materials physics and other applied sub-areas, makes up some 40 per cent of the total.

Other physicists explore the universe in the pure spirit of the Greek philosophers, aiming only to understand nature. Unlike industrial physics research applied to corporate goals, this is non-commercial research carried out at universities, or installations like CERN and LIGO. University researchers and national or international laboratories receive government funding, some of it at massive levels for exotic equipment like elementary particle accelerators, telescopes for astrophysics research, and spacecraft to explore the solar system and beyond (in the US, companies like SpaceX are beginning to privatize the government involvement in space technology). Theoretical work does not require hyper-expensive experimental apparatus, but typically needs high-level computation capability that brings its own costs.

Despite the apparent sharp distinction between 'pure' and 'applied' physics, they can connect in unexpected ways. A fundamental piece of evidence that the universe began in a 'Big Bang' is the cosmic microwave background (CMB), electromagnetic waves that originated in the early universe and fill all space. These were accidentally discovered using equipment built for commercial purposes. Conversely, the global positioning system (GPS) of space satellites uses the pure physics of general relativity to accurately determine real-time locations on the Earth's surface for civilian and military use.

Physics is what physicists do

Another way to define what physics is about, or at least to take its snapshot in a given era, is to look at what its practitioners are actually doing. However much physics may be defined in terms of its ideas and practices, it is also a living science defined by what interests and engages physicists at a given time.

But who or what is a physicist? Though the title goes back to the 19th century, there are still different definitions. A Bachelor's or Master's degree is sufficient background for some physics-related jobs but a PhD is necessary to conduct high-level research or teach university-level physics. Other criteria are publishing in physics journals or belonging to a physics society. That last indicates a professional commitment and has global weight because most advanced countries have such an organization: the American Physical Society (APS) in the US, the Institute of Physics in the UK with some 50,000 members, the German Physical Society (the world's biggest, with 62,000 members), and so on. With this as a criterion, a recent estimate is that there are 400,000 to 1,000,000 physicists worldwide, compared to about 1,100 physicists worldwide in 1900.

We can compare what physicists are doing now to what they were doing then. At the First International Congress of Physics in 1900, 60 per cent of the attendees had teaching positions, including university posts that also involved research; 20 per cent worked in industry, and 20 per cent in government. In 2014–16, various statistical sources for the US still showed up to 20 per cent in government but the academic percentage has dropped to no more than about 33 per cent, the difference being that the remainder is now employed in private sector research and development.

Physics research has evolved too. The keynote talks and contributed papers presented at the 1900 Congress covered classical mechanics, electricity and magnetism, light and optics, and thermodynamics along with the recent discovery of radioactivity, with a few papers on the properties of solids and other topics. Quantum theory and relativity theory had yet to be announced and the Congress had little to report at the large scale;

papers in the section called 'cosmic physics' mostly covered earthly and solar phenomena.

Today, judging by membership in the separate research divisions of the APS, about 20 per cent of physicists work in pure physics at the frontiers of the small and the large in nuclear and elementary particle physics, and astrophysics and gravitational physics—areas just being born in the early 20th century. The biggest single area is condensed matter physics, which garners 12 per cent of the membership and overlaps with materials physics at 6 per cent. The study of matter has grown far beyond its small role in 1900 to form a major part of basic research and in applications such as nanotechnology. Other areas where significant numbers of physicists carry out research are atomic, molecular, optical and laser physics. Even where these overlap with the topics covered at the 1900 Congress, they have been transformed by quantum theory.

Physics and life

Surveying the different definitions of physics shows the many facets of the science and how they change, but they all lie within the general definition 'the study of matter, energy, and the interaction between them'. Especially recently, however, the totality of physics has gone beyond its scope as defined in the *Oxford English Dictionary*: 'The branch of science concerned with the nature and properties of non-living matter and energy, in so far as they are not dealt with by chemistry or biology'. This is not quite complete, for physics has dealt with living biological systems for a long time and with chemical ones too.

Biological physics or biophysics was in evidence at the 1900 Physics Congress, which included five talks on the subject, and goes back at least to the 18th-century work of Luigi Galvani in what he called 'animal electricity'. Defined as investigations of 'the physical principles and mechanisms by which living organisms survive, adapt, and grow', and applying to all levels of life—from molecular and cellular to whole organisms and ecosystems—today it is a multidisciplinary area that overlaps with molecular biology and computational biology, and has its own professional societies and journals. Chemical physics too has a long history. It is the study of chemical systems from the viewpoint of the atoms that make them up. It was already known as a specific research area in 1900, and continues as a division of the APS and in dedicated journals.

These two blended areas show how the generality of physics in dealing with matter and energy allows it to mesh with other sciences, extending its definition and impact. We have not only biological physics and chemical physics but also astrophysics, which applies physical principles to interpret astronomical observations and probe the nature of celestial bodies; geophysics, the application of physical principles to understand the structure of the Earth along with its oceanic and atmospheric behaviour, with applications to the study of other planets as well; environmental physics, medical physics, and more.

Depending on how physics develops, a century from now its sub-areas and companion sciences might look different, or it could even be fundamentally redefined. If string theory or another theory were to successfully combine the Standard Model with general relativity, that would be the greatest correlation in physics yet; or if none of these approaches works we would be left either without a Theory of Everything—though physics would continue to progress even so—or possibly with the greatest upheaval in physics, the revolutionary option to bypass experimental verification.

Whatever that outcome, the current state of physics depends on how physicists in the past performed experiments and proposed theories, then interacted until a valid theory was hammered out, and how they continue to do so today.

Chapter 3
How physics works

After centuries of natural philosophy that became the science of physics, and especially in the last two centuries, physics has reached an enviable state. Today it is firmly grounded in classical physics, which accurately describes much of our immediate and relatively nearby world, the mid-range scale of the cosmos; and in modern physics, quantum mechanics and relativity, which describe much of the small and large scales of the universe that lie far beyond direct human reach.

A work in progress

These successes do not mean physics is stagnant. No contemporary physicist says 'we know all we need to about nature' or 'there are no new discoveries to be made' as some thought at the end of the 19th century. Perhaps having learned a lesson from that time, physicists today understand that physics still lacks important answers because of unexplained phenomena, because of new research tools, and because its aspirations, especially the quest for a Theory of Everything, have grown.

Among the unfinished work, there is the Standard Model that explains much but not all about elementary particles and omits gravity; quantum theory, which works well, but without our full understanding of its truly strange features; and the development of a theory of quantum gravity. Besides, many other questions, from the nature of dark energy and dark matter to what we still do not know even about ordinary matter, keep physics fresh and challenging.

How did physics reach this state? How did and still do physicists develop theories?—whether old incorrect ones, current ones we think are correct or mostly so, or theories-in-waiting that need confirmation. How did, and still do, other physicists decide what experiments they should perform to confirm or debunk theories or point towards new ones? And for both theory and experiment, how do we know they are 'correct'? Have physicists produced ideas and data that were just plain wrong, misleading, or badly interpreted, inadvertently or intentionally?

Above all, how do the two halves of the physics equation, theory and experiment, come together to produce truth, or 'truth' within the boundaries of physics? And how is physics research done in practice; who does it, and who pays for it? In short, how does physics work?

Basic beliefs

The answers are complicated. The content and intellectual style of physics have changed over the years, and individual researchers differ in how they choose and attack problems. Since the rise of classical physics and even earlier, though, some broad principles have applied to theory and experiment.

Two of these principles—or more accurately, philosophical positions—are implicit: that rational understanding of nature is possible, and that causes produce effects. Physics generally follows the rules for cause and effect put forth in 1738 by the Scottish philosopher David Hume: 'the cause and effect must be contiguous in space and time…the cause must be prior to the effect…The same cause always produces the same effect, and the same effect never arises but from the same cause', but has not always done so; for instance, in Newton's gravitational theory, bodies affect other non-contiguous bodies. These rules can also be prescriptive; one reason the theory of relativity bans communications that travel faster than the speed of light is that otherwise effects could precede causes, making time travel possible.

Another precept is useful in choosing among theories. Occam's (or Ockham's) Razor, from the 13th–14th century English philosopher William of Ockham, mandates that among competing hypotheses the simplest one with the fewest assumptions should be chosen. Besides aesthetic appeal, simplicity has a deeper value: a theory can always be made more complex to agree with data, as Ptolemy did with his geocentric astronomical model, but a theory stripped to bare essentials that still makes definite predictions can be more rigorously tested.

A third essential is faith in what the Nobel Laureate Eugene Wigner has called the 'unreasonable effectiveness of mathematics'. From early applications to today's mathematically sophisticated theories, we have found that the apparently abstract discipline of mathematics somehow maps onto the real world to reliably describe and predict its operations, often for phenomena beyond the original scope of the mathematics. Wigner notes that 'the enormous usefulness of mathematics in the natural sciences is something bordering on the mysterious' but this inexplicable usefulness is crucial for physics.

More specific to physics are its general conservation laws for properties that do not change over time in an isolated physical system. Besides the conservation of mass and energy, other conserved quantities are linear momentum, the product of mass and velocity for each element of a system such as two colliding billiard balls; angular momentum, for rotating systems like the planets in their orbits; and electric charge. Various properties of elementary particles are also conserved. These laws are absolute and constitute some of the great truths of physics. A theory or experimental result that violates them would be rejected, unless it incorporates a loophole that makes the violation only apparent, which can happen at the quantum level.

Motivation

Within these general guidelines, new theories and experiments arise from specific motives besides scientific curiosity. One goal for a theorist is to unite a set of data with a theory that accurately describes and predicts how some part of nature operates, as Maxwell did with electromagnetism, and as many theorists did to produce quantum physics and the Standard Model. Other goals may be to improve or extend an older theory, as when Copernicus put the sun instead of the Earth at the centre of the universe.

A crucial motivation comes when an established theory fails to replicate data, as when Max Planck had to invent the quantum to explain the observed pattern of electromagnetic radiation from a hot source; or when a theoretical approach raises internal questions or reaches an apparent dead end.

Experimentalists have varied interests as well. Some specialize in accurate measurements of important quantities, as Albert Michelson did for the speed of light. Others design experiments to test existing ideas, as Ibn al-Haytham did to determine if light originates from the human eye or from external sources. Some are inspired to make observations or carry out experiments to verify—or not—new theories, which is not always easy. It took whole scientific expeditions to confirm Einstein's theory of general relativity, and it requires incredibly delicate experiments to confirm certain quantum effects.

Least predictable is the role of serendipity in physics discoveries, which besides luck requires alert and curious experimenters who see an anomaly and follow it up. Fortunately these were on hand to discover X-rays in 1895 and the CMB in 1964. Serendipity happens in theoretical work too. In 1961, the American meteorologist Edwin Lorenz noticed that entering 0.506 instead of 0.506127 for the starting value of a variable in his computer-generated weather predictions radically changed two months of simulated weather. That began Lorenz's contributions to chaos theory, the study of systems whose dynamic behaviour depends strongly on starting conditions, summarized by Lorenz as the 'butterfly effect', the tiny flap of a butterfly's wings that might eventually produce a tornado.

Experiment becomes important

The relationship between theory and experiment has changed as physics has developed. Aristotle and other early natural philosophers asserted propositions about the real world that might fit into their overall intellectual schemes but were not backed up by data. When later in the history of physics new theories were put forward, their validity did not stand solely on assertions or even on mathematical consistency: they were expected to organize or explain known data and to make predictions that could be confirmed by experiment.

Once established, the experimental approach took different forms. Some breakthroughs came from just one or perhaps two researchers such as Galileo for mechanics, and Newton for light, in the early days of physics; Thomas Young, and Michelson and Morley, in the 19th century; and many in the 20th such as the American physicist Theodore Maiman who invented the first working laser in 1960. At the First International Congress of Physics in 1900, the experimental presentations had only a single author except for a handful with two (including a talk by Pierre and Marie Curie). This scale of small-group experimentation with 'table-top' equipment continues; for instance, the 2010 Nobel

Prize in Physics went to two University of Manchester researchers for their work on graphene, a two-dimensional form of carbon.

Other experiments are carried out by teams of up to thousands of physicists at a given installation or linked globally by the Internet, using huge machines like accelerators and gravitational wave detectors with construction and operational costs of billions of dollars. These 'big physics' efforts along with table-top ones are funded within government laboratories or by grants to academic researchers from governmental agencies, such as NASA, the Department of Defense (DoD), and the National Science Foundation (NSF) in the US. Governments do not necessarily value physics as a key to fundamental knowledge, but they see its importance in supporting applications from consumer and medical technology to military use, so enhancing economic growth, citizen well-being, and national security.

The boundaries between table-top and big physics can be fluid. My own career as an experimentalist has involved small groups of students and colleagues using lab-size lasers and spectrometers, and other efforts at huge installations like the Los Alamos National Laboratory in New Mexico, and the Brookhaven National Laboratory on Long Island, New York, where my research used a synchrotron, a powerful ring-shaped light source nearly a kilometre around.

In contrast to the need for physical facilities and research teams, theorists can develop ideas or entire theories pretty much on their own. Electromagnetic theory, relativity, quantum theory, and string theory or their seeds came from individuals who needed only time to think and talk to colleagues, access to data and other theories, mathematical proficiency, and pencil and paper, or later, a calculator or computer. Other theorists might then refine and expand the original idea while experimentalists work to confirm it.

Today mixed groups of experimentalists and theorists operate at research centres like national laboratories and universities; for example, the physics faculty at the Massachusetts Institute of Technology (MIT) includes twenty-six experimentalists and twenty theorists in nuclear and particle physics. This encourages the creative day-to-day interchange of ideas besides what occurs more formally in physics conferences and through research publications. Gathering data and working out its implications can take years and the efforts of many researchers before the results become established physics wisdom, the process that formed the Standard Model.

Theorists and experimentalists use different skills. All physicists need mathematical ability but theorists must master deeper levels of mathematics as a tool; Einstein needed highly abstract mathematics to develop general relativity. Experimentalists may use commercial equipment such as lasers; but for novel experiments they must know how to design and often themselves build specialized apparatus, such as the massive installations that detect elementary particles at CERN, and must know how to analyse data and judge its reliability. Both types of physicist, however, use similar personal qualities of creativity, imagination, and intuitive insight into physics to choose and pursue significant research problems.

These abilities may show up during undergraduate study but the committed choice of theory or experiment and the honing of expertise and physical intuition happens in graduate work. After earning a Bachelor's degree, physicists who want to do original research invest another six years on average to finish a dissertation and earn a doctor of philosophy (PhD) degree under an adviser. Often they go on to postdoctoral work in a research group before starting an independent career.

In the US nearly 1,800 people earned physics PhDs to qualify as researchers in 2012, a number that has trended upwards for over a century and has come to include more international recipients and more women. Only a few women participated in 19th- and 20th-century physics; earlier figures such as Marie Curie, the theorist Emmy Noether, and the astronomer Henrietta Leavitt have been followed by more recent figures such as the astronomers Jocelyn Bell and the late Vera Rubin, and the late condensed matter physicist Mildred Dresselhaus. Even by 1983, only 7 per cent of US physics PhDs went to women. By 2012 that had nearly tripled to 20 per cent, leaving women still a distinct but growing minority in the 21st century (and in this century, the Canadian laser physicist Donna Strickland shared in the 2018 Nobel Prize in Physics).

The flavour of the interaction between experimentalists and theorists, and experiment and theory, appears in case studies of some high points in physics. These show how the two areas arose and now influence each other in different ways, from painstaking analysis of data to accidental discoveries and moments of inspiration.

Measuring the Earth with a shadow and a stick

Extending natural philosophy to the realm of physical measurements, around 240 BCE, the scholar and mathematician Eratosthenes, chief librarian of the famous Library of Alexandria, obtained an early geophysical result when he determined the circumference of the Earth. He noted that on the date of the summer solstice in Syene, Egypt (now the city of Aswan), the shadow of a person looking into a deep well blocked the reflection of the sun, which meant it was directly overhead. At the same date and time, in Alexandria far north of Syene, when he measured the length of the shadow of an upright stick cast by the sun, he found that the sun appeared displaced 7.2 degrees from the vertical, 1/50 of a full circle of 360 degrees (Figure 5). Therefore the known distance between the two cities was 1/50 of the Earth's circumference.

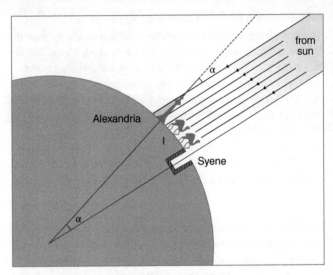

5. The Greek scholar Eratosthenes determined the Earth's circumference c.240 BCE.

When Eratosthenes did the mathematics, his result for the circumference—allowing for uncertainty in converting the units he used into modern units—was remarkably close to the correct value, 40,000 kilometres. This foray into actual measurement showed the power of careful observation, experimental data and its analysis, and also that underlying assumptions matter. Eratosthenes, after all, had to start with the premise that the Earth is round.

Data, modelling and gravitational theory 1.0

The astronomical models that Ptolemy, Copernicus, and Kepler constructed also depended on data in the form of astronomical observations, and on the assumptions built into the models. In the 2nd century CE Ptolemy's model incorporated the belief that the heavenly bodies travel around the Earth in perfect circles. To make this agree with all the observed behaviour of the planets such as retrograde motion—their apparent backwards movement at certain times—each planet was postulated to move in a small circle called an epicycle whose centre moved around the Earth in a large circle called the deferent, and the Earth was relocated from the exact centre of the heavens.

The Ptolemaic model was the standard for predicting astronomical behaviour for over a millennium until Copernicus improved it in 1543 by putting the sun, not the Earth, at the centre of the universe. This neatly explained retrograde motion as due to the movement of the Earth relative to the other planets and put the planets in correct order of distance from the sun. But the model continued to take planetary orbits as circles so it still required epicycles and was not more accurate than Ptolemy's, until Kepler provided the final adjustment. After carrying out hundreds of pages of hand-written calculations based on Tycho Brahe's accurate observational data for the position of Mars, Kepler concluded that the planets move in elliptical orbits and obey two other laws of planetary motion.

Then Newton changed physics by relating these three laws to a single physical cause. His theory of universal gravitation summarized Kepler's results in an equation for an attractive force acting between the sun and a planet—or between any two bodies—along the line joining them and varying inversely with the square of the distance between them. This simplification followed Newton's own paraphrase of Occam's Razor: 'We are to admit no more causes of natural things than such as are both true and sufficient to explain their appearances'.

Newton's theory accurately described planetary motion but needed to be tested in the laboratory as well, as was first done in 1797 by the English physicist Henry Cavendish. His clever arrangement put a small lead mass at each end of a rod some 2 metres long that was suspended from a wire. Near each of the masses, on opposite sides of the rod, he put a 158 kilogram piece

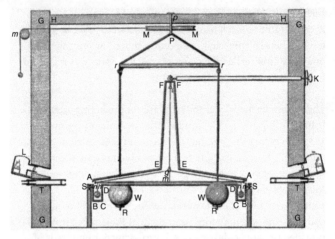

6. Henry Cavendish measured gravitational effects in 1797 to test Newton's theory.

of lead (Figure 6). As Newton predicted, gravity acted between the big and small masses, producing a turning force that made the rod rotate until the wire exerted an equal and opposite twist. From this Cavendish found the gravitational force between the masses and determined a value for G, the gravitational constant in Newton's theory, within 1 per cent of the modern value.

A *Gedankenexperiment* and gravitational theory 2.0

This might seem to end the search for a theory of celestial motion but Newton's approach had problems. Though gravity as a force was a natural idea within his mechanistic universe, it brings up the troublesome issue of 'action at a distance', that is, non-contiguous cause and effect as objects somehow influence each other without physical contact. Newton himself found this unsatisfactory. Another problem is that Newton's theory requires that gravitational effects travel instantaneously between bodies, but special relativity forbids travel faster than light.

Albert Einstein resolved these issues in 1915 by treating gravity differently. His theory of general relativity sprang out of Einstein's unique creative approach, a *Gedankenexperiment* ('thought experiment' in his native German). He has related how in 1907 he visualized a scene that at first startled him and that he later called his 'happiest thought:'

> I was sitting in a chair in the patent office at Bern when all of a sudden a thought occurred to me. If a person falls freely, he will not feel his own weight.

This apparently simple insight turned out to be profound, for from it Einstein reasoned his way to the view that gravity depends on the geometry of spacetime, the four-dimensional entity he had derived in special relativity. In general relativity, a big mass like a star distorts spacetime, changing the straight line path an object would follow in empty space into the orbital motion of a planet and all other gravitational effects. The American theorist John Wheeler summed this up in a pithy epigram: 'Spacetime tells matter how to move; matter tells spacetime how to curve'.

The theory was validated when the English astrophysicist Arthur Eddington mounted an expedition to the island of Príncipe near Africa, and Greenwich Observatory sent one to Brazil, to view the total solar eclipse of 1919. There the temporary darkness made it possible to confirm Einstein's prediction that light from a distant star would bend by a certain amount around a massive object like our sun. This result created a huge splash in the press and made Einstein world-famous.

Other experiments confirm that the flow of time changes in a gravitational field as the theory predicts. General relativity also accounts for discrepancies in the orbit of the planet Mercury, and in 1916, it was used to predict the possibility of black holes, which have since been found. Its last major prediction, that massive cosmic events like a collision between black holes produce gravitational waves travelling at the speed of light, was confirmed by LIGO in 2015.

Do these great successes mean Newton's theory is wrong? No; it works well as a limiting case of general relativity for bodies moving far below the speed of light or not too near a star, and it is mathematically simple. General relativity has been called the most beautiful physics theory because of its elegant idea that gravity comes from spacetime, the very fabric of the universe. But the theory uses ten complicated nonlinear equations that can be fully solved only by computer, satisfying Occam's Razor in conceptual but not mathematical simplicity. A more serious issue is that its geometric nature makes the theory unlike any other in physics including the Standard Model, one reason that merging the two is difficult.

Small scale serendipity

In contrast to the long search that led to a theory of gravitation, requiring much poring over data and succeeding improvements to the theory, other important findings in physics have been sudden and unexpected. These might not have been appreciated without an approach that brings both curiosity and scientific thoroughness to the study of nature; as the great French biologist and chemist Louis Pasteur put it in 1854, 'In the fields of observation chance favours only the prepared mind'.

One example is Röntgen's discovery of X-rays in 1895, which won him a Nobel Prize. He noticed that when he turned on a Crookes tube in his lab, a fluorescent screen 3 metres away began to glow though the tube was covered with opaque material. Röntgen found that the glow was not due to cathode rays but to an unknown type of penetrating radiation, later shown to be very short wavelength electromagnetic waves. One early X-ray image showed the bones in his wife's hand, a remarkable result soon put to use for medical imaging (Figure 7).

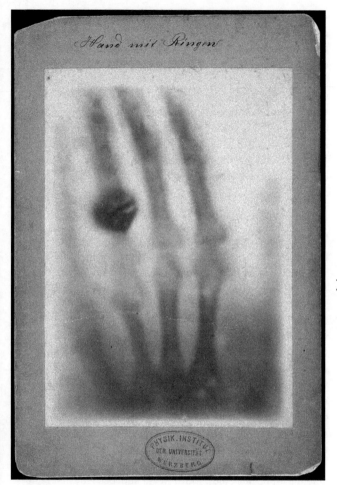

7. Soon after discovering X-rays in 1895, Wilhelm Röntgen displayed the bones in his wife's hand.

Röntgen's sensational accident led to another. The French physicist Henri Becquerel had been studying phosphorescent uranium compounds, theorizing that they might absorb sunlight and re-emit it as X-rays. In 1896, he put the compounds and some photographic plates into a dark drawer and was later surprised to find that the plates had been exposed without a light source. Further work showed that the uranium compounds and pure uranium emit X-rays and other radiation entirely spontaneously. Becquerel had discovered radioactivity, and shared the 1903 Nobel Prize in physics with Marie and Pierre Curie who continued the research.

Large scale serendipity and cosmic origins

These accidental discoveries at the small scale were matched by two at the large scale of the universe. Intriguingly, both had roots in engineering projects carried out at Bell Telephone Laboratories, Holmdel, New Jersey for commercial rather than research purposes.

In 1928, the American physicist Karl Jansky, working at Bell Labs, was seeking sources of static that would interfere with a new transatlantic radio telephone service. His antenna at Holmdel detected static from thunderstorms and also registered a signal from no known source. Tracking the signal over time and in different directions, he concluded that it came from near the centre of our own Milky Way galaxy. His paper in 1933 about the first radio signals from beyond the Earth, 'Electrical disturbances apparently of extraterrestrial origin', received wide attention. Bell Labs rejected his proposal to build a bigger dish-style antenna for further study, which would have been the first radio astronomy observatory, but radio astronomy went on to flourish.

Three decades later, Bell Labs radio astronomers Arno Penzias and Robert Wilson made another fortuitous discovery not far from the site of Jansky's antenna. In 1964, while detecting radio waves from space using a horn-shaped antenna 6 metres tall, originally built to test signals from communications satellites, they found an unexpected signal at the microwave wavelength of 7.35 centimetres. It did not come from known extraterrestrial sources or from terrestrial ones such as nearby New York City, nor was it an artifact due to pigeon droppings on the antenna. It was a real signal, pervasive and unchanged no matter where in space they pointed the antenna, and it needed an explanation.

At the time there were two different ideas about the origin of the universe, which was known to be expanding as observed by Edwin Hubble in 1929. In the steady state theory, the universe remains homogenous and unchanging as it expands because new matter is being continuously created. In the Big Bang theory, the universe began from a single point and extremely high temperatures and expanded to reach its present state. Penzias and Wilson found that the American physicist Robert Dicke had calculated that a Big Bang would leave a residue of space-filling electromagnetic waves, the cosmic microwave background or CMB. This oldest radiation in the universe would be of the type called blackbody radiation.

Blackbody radiation is emitted by the vibrating atoms in any object whose temperature is above absolute zero, with an intensity and spectrum that depend on the temperature. Our sun emits this radiation at about 5,800 kelvin (5,500 degrees Celsius), much of it at the wavelengths we see, 400 nanometres to 750 nanometres. Dicke had predicted that space would be filled with microwave blackbody radiation originating at around 3 kelvin (−270 degrees Celsius, nearly at absolute zero), the temperature of space after the universe cooled down from billions of degrees. When the initial measurement by Penzias and Wilson was later extended over the whole microwave range, the data exactly matched theory for a blackbody at 2.7 kelvin. This spectacularly good agreement was strong evidence for the Big Bang, now the accepted theory.

A radical quantum theory

Long before the time of Penzias and Wilson, physicists in the late 19th century tried to develop a classical physics theory of blackbody radiation but their efforts disagreed with experimental spectra. After wrestling with the problem, the German physicist Max Planck introduced something new by assuming that the total energy of the vibrating bodies generating the radiation was composed of many extremely small and separate units of energy. This led to a new equation that Planck announced in 1900. It fitted experimental data perfectly and is still used for blackbody radiation calculations including the CMB.

Planck thought his assumption was just a mathematical trick, but he was doing something profound by imagining separate packets of energy, or quanta. Without fully grasping it this 'reluctant revolutionary', as the science historian Helge Kragh calls him, was laying the roots for quantum mechanics, the theory of the small and one of the two great disruptive theories of 20th-century physics. Few of his peers immediately saw the implications and quantum theory did not receive much attention until the 1911 Solvay Conference with the theme 'On radiation theory and the quanta'.

However, Einstein was an early adopter. In 1905, he created the idea of quantized packets of light energy, later called photons, to explain the photoelectric effect where impinging light knocks electrons out of a metal plate, and earned a Nobel Prize. This introduced the puzzling quantum wave–particle duality where light behaves both as a wave and a particle-like photon. In 1924 the French theorist Louis de Broglie surmised that the same duality occurs for matter. Soon after, the American experimentalists Clinton Davisson and Lester Germer, working at Bell Labs, showed that electrons can scatter from the atoms in a crystal just as waves do. In 1928 other experiments showed that light interacts with an electron like one billiard ball hitting another, further establishing the reality of photons. Both light and matter, it seemed, could be both wave and particle.

Meanwhile theorists further explored quantum physics and the wave nature of matter. In 1926, the Austrian theorist Erwin Schrödinger derived an equation to describe quantum systems. Instead of using Newton's law $F = ma$, Schrödinger's equation describes the motion of 'matter waves' in a way that conserves energy. This yields the mathematical 'wave function' from which one can calculate the position, energy, or any other property of a quantum object such as an electron in an atom. But these can only be given as probabilities; it was no longer possible, for instance, to put an electron at a definite location in space but only to give a range of possibilities.

This 'fuzziness' in our knowledge also shows up in the Uncertainty Principle. As derived in 1927 by the German physicist Werner Heisenberg, this states that in the quantum world it is impossible to simultaneously know some quantities such as the position and momentum of an electron with absolute precision. Einstein was unhappy with this indefiniteness, believing that there is an underlying determinism in nature. Other aspects of quantum theory also remain perplexing. Nevertheless, the theory works well. By the 1930s, the Schrödinger equation was being successfully used to explain the properties of atoms, of metals, and of semiconductors, the materials that would come to underlie digital electronics and the computer age.

But the Schrödinger equation was not complete because it omitted special relativity. In 1928, the British theorist Paul Dirac derived a new equation that combined quantum theory and special relativity to describe electrons moving near the speed of light. This gave an unexpected result, implying that for every elementary particle such as an electron there exists an 'antiparticle' that is identical except with opposite electric charge. Dirac's astonishing prediction was confirmed in 1932 with the discovery of the positron, the antiparticle to the electron. Other antiparticles have been found since and now we know that antimatter is part of the universe. Matter and antimatter annihilate each other in a flash of energy when they meet, but such eruptions are not often seen because there is little antimatter in the universe.

Dirac also made initial steps towards a quantum field theory of electromagnetism that includes relativity, to deal with photons travelling at the speed of light. After considerable further effort by many contributors, in 1948 the American theorists Richard Feynman and Julian Schwinger, and the Japanese theorist Shin'ichiro Tomonaga finally developed quantum electrodynamics (QED), the quantum theory of electromagnetism. It shows how light and matter interact by exchanging photons and was confirmed through its extremely accurate numerical predictions for the energy levels of a hydrogen atom.

QED explained electromagnetism, one of the four fundamental forces in the universe. Further work showed that two of the other three forces also come from the exchange of elementary particles. The strong nuclear force, which holds quarks together to make protons and neutrons and binds these into atomic nuclei, arises from the interchange of massless particles called gluons. The weak nuclear force, which appears in radioactive decay, arises when neutrons, electrons, and other particles exchange any of three types of particles called W^+, W^- and Z bosons. (Electromagnetism and the weak force are now known to be aspects of a unified 'electroweak' force.)

After much effort since the mid-20th century, all these results evolved into the Standard Model, which organizes all the known elementary particles into groups (Figure 8). Besides photons, gluons, W^+, W^- and Z bosons—the five so-called 'gauge boson' particles that carry forces—the model includes six quarks and six leptons that make up matter. The theory was experimentally

Standard Model of Elementary Particles

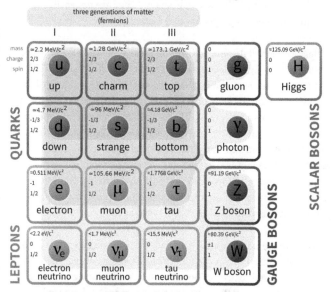

8. **Efforts by many physicists led to the Standard Model, the quantum field theory of fundamental particles.**

verified when the particles it predicted were detected; the first two types of quarks in 1968, followed by the remaining four types of quarks, other elementary particles, and finally the Higgs boson (also called a 'scalar boson') in 2012.

Strings and the multiverse

The Standard Model is a triumph of modern physics but it is incomplete. It does not predict all the properties of elementary particles or include dark matter. Most frustrating, it has so far proven impossible to combine it with general relativity to incorporate gravity. The quantum world is discontinuous whereas spacetime as it appears in general relativity is smooth, and it is difficult to conceive that it may have a discontinuous nature. Each theory is valid in its own arena but they are not easily combined into a single theory of quantum gravity.

The approach that some believe will resolve the impasse is string theory. Drawing on decades of theoretical work, its central idea is to replace the accepted view of elementary particles as point-like masses with one-dimensional entities called strings. Like a guitar string, these can vibrate in different modes that correspond to all the known elementary particles and introduce a brand new one, the graviton—the speculative particle that would cause gravity just as photons, gluons, and W and Z bosons cause the other fundamental forces. This would make string theory the long-sought theory of quantum gravity and a possibility for a Theory of Everything.

Having the graviton appear naturally out of string theory would be deeply satisfying but the theory has problems. It has yet to produce a testable prediction and will be difficult or impossible to verify by experiment. The strings at its heart are 10^{-35} metres long, far too small to observe by any means we can conceive. The theory also requires more than the four dimensions of spacetime—up to eleven dimensions, seven of which are hidden because they are 'compactified', curled up to become too tiny to discern in normal experience or by any experiment within our grasp.

There is also a complicating structural issue. The extra dimensions in the theory can be configured in up to 10^{500} different ways, corresponding to that many universes with diverse physical properties that together comprise the so-called multiverse. But it is highly questionable whether we can observe these supposed other universes, and their variety may make it impossible to obtain definite predictions. The American theorist Paul Steinhardt has called the multiverse a 'Theory of Anything' that is of little value because it 'does not rule out any possibility' and 'submits to no do-or-die tests'.

Because the mathematical structure that string theorists have built seems unlikely to be confirmed by experiment, many physicists dismiss it. But its champions think it is probably correct. They suggest that the theory should be accepted without experimental verification, and the philosopher of physics Richard Dawid has proposed that physicists should consider entering an era of 'post-empirical science'. Others, such as the theorist George Ellis and astrophysicist Joe Silk, find this idea deeply harmful to the integrity of physics; and the German theorist and blogger Sabine Hossenfelder has written that the very idea of 'post-empirical science' is a contradiction in terms. But the LQG theory that quantizes spacetime instead of postulating the existence of strings is an alternative approach, and projected measurements of quantum effects in space may provide the stimulus of new data.

The human factor

The fact that physicists have different opinions about the value of string theory has an important sub-text. It reminds us that physics is done by people, whose personal qualities and individual approaches matter along with their scientific insight. Despite their training, as in any human activity, physics researchers are subject to error and are propelled by personal as well as scientific motivations—the desire for success and recognition; the powerful impulse to 'scoop' a competitor or prove one's own ideas right; and the desire for promotion, academic tenure, and research funding, which are not easy to obtain.

Personal motivations typically enhance the drive and commitment necessary to do good research, but if these motivations overcome sound judgement and lead to poor, misleading, or even fraudulent work, then, like science in general, physics needs to be self-correcting. Unverifiable results should be and have been caught by peer reviewers or other researchers. Examples of research gone wrong like the supposed observation in 1989 of cold fusion—the generation of energy from hydrogen nuclei merging at ordinary temperatures rather than the hundreds of millions of degrees known to be necessary—and its swift debunking are signs that physics can fix itself when needed.

Subjective motivations, not objective scientific quality, can also influence the research problems physicists choose. The British theorist Roger Penrose has written about the 'fashionable' physics of an era, meaning a focus on a single strategy for dealing with a given problem without properly evaluating its validity. This happened in the past, he argues, with the wide acceptance of the Ptolemaic model and has happened with string theory, where a 'bandwagon' effect may cause researchers to fear they will be sidelined unless they follow the prevailing fashion.

Changing how physics works?

The correctness of a theory is not decided by popular vote or by counting how many people are working on it, but by whether it agrees with experiment, an approach that began when Galileo's experiments in mechanics showed the importance of empirical data. For nearly 400 years that has been a successful model but the difficulties with quantum gravity have led some to question the empirical approach. This may lead to a critical moment, but we should remember that physics weathered the twin revolutions of relativity and quantum physics, and emerged stronger. Whether quantum gravity and a Theory of Everything remain forever out of reach, or answers finally appear, the search for them deepens our understanding of nature.

In any case, much of today's physics operates differently. Rather than concentrating on general theories that explain the universe, these efforts extend into applications and into related scientific areas that successfully explore the universe and deeply affect our own lives.

Chapter 4
Physics applied and extended

Physics is what physicists do, I wrote earlier. By that criterion, the Greek natural philosophers whose ideas led to physics might be surprised at what some of today's physicists are up to and what that means for physics. Apart from the 20 per cent of researchers who study the pure physics of quantum theory and relativity, the numbers I cited show that many physicists work in industrial and applied physics or in interdisciplinary areas like astrophysics and geophysics.

These connections arise because foundational physical concepts and theories like energy and quantum mechanics apply broadly across science, and serve as a basis for technology and its industrial use. What is particularly striking about partner areas like medical physics and environmental physics is that they directly affect daily life and society in ways that pure physics does not, but still depend on basic research as a foundation for their important applications.

Instruments and theories

These applications and connections come about in different ways. One route is through instruments and processes that use physical methods, or were created or discovered in physics labs but have wide uses outside them. X-ray imaging is a prime example.

It revolutionized medical practice soon after it was discovered, and has since been supplemented with MRI, ultrasound imaging, and other techniques based on physical principles that can examine and help heal the human body.

Another kind of connection arises when physical theories provide tools or a conceptual framework that supports other sciences or technology. Newton's explanation of ocean tides as due to the gravitational effects of the moon and sun is essential for geoscience; quantum theory is necessary for applications of light and lasers, and for nanotechnology; and the complex theory that describes the motion of liquids and gases underpins meteorological forecasting and climate modelling.

Some of these connections are long-standing. Astrophysics, geophysics, and medical imaging technology go back to earlier days. Others have arisen or gained new momentum in modern times. As biology moves to the molecular level and adopts quantitative approaches and physical techniques, research in biological physics has been invigorated. Environmental physics, a new field that is growing along with concerns about the human impact on the Earth, contributes by providing methods for the clean and efficient generation and use of energy. Nanotechnology and quantum computing are other emerging areas that depend deeply on physics, but the older interdisciplinary areas remain vibrant as well.

To the stars

Though astrophysics originated in the ancient science of astronomy, today it is an especially active blended area. It uses physics tools and principles to extend classic observational astronomy, which has for centuries tracked heavenly bodies. Applied physics broadens astronomical observation in the design of advanced optical telescopes and other telescopes and detectors for radio waves, microwaves, infrared and ultraviolet light, X-rays, and gamma rays—along with cosmic rays, elementary particles that come from deep space. Physics applied to rocketry and space technology makes it possible to put some of these observational platforms such as the Hubble Telescope into Earth orbit to avoid atmospheric interference.

Physical methods also serve to analyse data gathered by these devices to determine the dynamics and composition of planets, stars, galaxies, and the universe itself, from the CMB to the interstellar medium, dark matter, and dark energy. Further, we need relativity and the Big Bang theory to interpret these results.

One essential astrophysical tool is spectroscopy, the study of how matter emits and absorbs electromagnetic waves and other radiation. Newton performed an early spectroscopic experiment when he passed sunlight through a glass prism and saw a continuous band of colour varying from red to violet, the sun's blackbody spectrum spread out by wavelength. The German physicist Joseph Fraunhofer refined the method when he invented the spectroscope in 1814. This combination of a prism and a lens broke sunlight into its wavelengths at a higher resolution than Newton achieved, and showed many narrow dark lines overlaid on the colours of sunlight.

These lines were later shown to represent the composition of the source. A hot gas emits energy at definite wavelengths that come from quantized atomic transitions in the gas and provide a unique fingerprint for any atomic element such as hydrogen. Elements can also be identified in absorption spectra, when quantum transitions in the cooler outer layers of a star absorb radiation from its hotter interior and appear as dark lines where energy is missing at the characteristic wavelengths (Figure 9). Emission and absorption spectra tell us the make-up of astronomical bodies and have produced surprises. In 1868, a previously unknown

9. Quantum transitions produce characteristic spectral lines that show the composition of celestial bodies.

emission line at 587.49 nanometres was observed in our sun's spectrum. It was a signature of the element helium, a major component of active stars found only later on Earth.

Spectroscopic data also reflect celestial motion. In the Doppler effect, light from a moving object is shifted towards longer red or shorter blue wavelengths depending on whether the body is moving away from or towards the observer, by an amount that depends on the speed of the body. In 1929, Edwin Hubble noted the red-shifted light from galaxies he observed through the Mt Wilson telescope (see Figure 2), showing that they are receding from us and, it turned out, also receding from each other. Hubble's result was the first definitive observation that we live in an expanding universe.

Many other examples show the power of physics ideas in astrophysics and, conversely, the power of astrophysical observations to answer physics questions. For instance, the detection of gravitational waves by LIGO in 2015 is an important physics result.

Around and inside the Earth

Physics contributes closer to home as well through geophysics, part of the collection of related areas that includes geology, meteorology, oceanography, seismology, and terrestrial magnetism that examine and analyse our own planet and its phenomena. One example is earthquakes, whose study has a long history going back to the early Chinese. By the 19th century, physicists had designed instruments to measure and record the ground motion caused by earthquakes as the seismic waves they create travel through the Earth. Then researchers found that analysing these disturbances with the physical theory of wave behaviour in a medium could yield information about our planet's interior.

Using this approach, in 1906 the British geologist Richard Dixon Oldham showed that the Earth has a central core. Later study found a boundary between the core and the layer above it, the mantle, 2,900 kilometres below the surface. The analysis was extended in 1936, when the Danish geophysicist Inge Lehmann showed that the core has a solid inner part surrounded by a molten outer part; and in 1996, when the American geophysicists Xiadong Song and Paul Richards showed that the inner core rotates slightly faster than the rest of the Earth (Figure 10). Now earthquake analysis continues along with new methods to probe our planet, as in the Rotational Motions in Seismology (ROMY) installation near Munich, Germany, that will use lasers to study the Earth's structure, and in the search for mineral resources such as petroleum.

Physical methods also yield the age of the Earth. This had been much disputed in the late 19th century, as methods like thermal

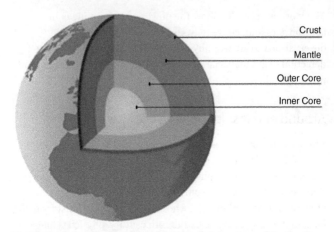

10. Much of what we know about the internal structure of the Earth comes from geophysical analysis of earthquake waves.

analysis gave values up to hundreds of millions of years rather than the biblical values of thousands of years. A more definitive approach, radioactive dating, began in 1907 when the American chemist and physicist Bertram Boltwood showed that uranium decays into lead. He measured the ratio of lead to uranium in rocks and obtained ages for the Earth up to 2.2 billion years. Finally in 1930 the British geologist Arthur Holmes refined radioactive dating to give a reliable geological clock, which today yields an age of 4.54 billion years for the Earth.

Another important application of physical ideas is in studying the Earth's climate and human-induced global warming. The basic mechanism, the Greenhouse Effect where atmospheric gases trap the sun's heat, was recognized by the French mathematician and physicist Joseph Fourier in the 1820s. In 1856, the American scientist Eunice Foote first showed that carbon dioxide (CO_2) effectively traps heat. Then John Tyndall further explored CO_2 and water vapour as heat-trapping agents, and the Swedish Nobel Laureate physical chemist Svante Arrhenius found a relation between the atmospheric concentration of CO_2 and global temperatures. These results and later research inspired today's understanding that human-caused processes are increasing CO_2 levels and raising global temperatures to harmful levels. The climate modelling that predicts the increase depends on theoretical and empirical studies of how solar radiation is absorbed, reflected and trapped to change the Earth's temperature.

Inside the body

Like the science of climate change, medical physics has 19th-century roots dating back to the discovery of X-rays and their ability to probe within the human body. Similarly, early research in radioactivity quickly led to the medical use of this new phenomenon. These threads continue in today's medical physics in various forms of imaging, in radiation therapy, and nuclear medicine as well as more recent approaches such as laser surgery and nanomedicine.

X-ray imaging remains an important diagnostic technique and has been further developed into computerized axial tomography (CAT) scanning, where many X-ray images of a patient are taken from different angles. A computer assembles these into a representation that shows soft tissues such as the brain and abdominal organs, which do not show up well in conventional X-ray pictures. One drawback of X-ray imaging is that these rays carry enough energy to ionize atoms and molecules by freeing their electrons and so can damage DNA. For this reason, modern techniques minimize patient exposure as much as possible.

However, X-rays can also kill cancer cells. The field of radiotherapy was initiated in 1900 when X-rays were used to treat skin cancers, and later, deeply buried tumours. Radiotherapy was extended with Marie Curie's discovery of radium in 1898. Its gamma rays were found to cause burn-like lesions to the skin like those from X-rays and radium therapy was quickly applied to cancer. Radium therapy was also considered something of a wonder cure for other ills—unfortunately without recognizing that gamma rays injure healthy cells, until the long-term harm of these rays was established in the 1930s. Madame Curie herself died in 1934 suffering from cataracts and anaemia or leukaemia, probably from years of radiation exposure.

Even where radium was properly used, it was rare and expensive, with only 50 grams available worldwide for radiotherapy in 1937. In 1935, the married couple Irène Joliot-Curie (daughter of Pierre and Marie Curie) and Frédéric Joliot won the Nobel Prize in Chemistry for discovering a substitute—artificial radioactive isotopes made by bombarding light elements like boron with helium nuclei. The medical value of these artificial isotopes was immediately apparent as expressed in the Nobel award speech to the recipients:

> The results of your researches are of capital importance for pure science, but in addition, physiologists, doctors, and the whole of suffering humanity hope to gain from your discoveries, remedies of inestimable value.

After World War II, nuclear reactors were used to create radioactive materials like ^{60}Co, the cobalt isotope of atomic weight 60, for radiotherapy. Then in the 1980s, the linear particle accelerator or LINAC offered an alternative. In a LINAC, electric fields bring electrons, protons, or ions to high speeds for research purposes. Medical LINACs accelerate electron beams that are used directly for radiotherapy or guided to smash into heavy metal targets to produce high-energy X-rays. Coupled with CAT imaging, these beams can be precisely targeted for maximum benefit and minimum undesirable effects.

Radioactive isotopes are also used in positron-emission tomography (PET), a practical application of antimatter that detects metabolic activity in the body. The patient is injected with a biologically active compound, typically a special formulation of the sugar glucose, that includes a weakly radioactive isotope. The isotope decays by emitting positrons, each of which travels a short distance in the body before encountering its antiparticle, an electron. They mutually annihilate, giving a burst of energy in the form of two gamma rays that travel in nearly opposite directions. These are detected externally and computer analysis tracks back their paths to determine their point of origin in the body.

After enough such events are analysed, the result is a three-dimensional image of regions where there is high metabolic activity. PET is most often used to display tumours and track the spread of cancer, but other applications include imaging the brain to confirm diagnosis of Alzheimer's disease. The method may also show promise to detect chronic traumatic encephalopathy, the degenerative brain condition that can appear in football players after repeated head trauma.

Physics provides other imaging methods as well. Medical ultrasound had origins in World War I when the French physicist Paul Langevin used underwater sound waves to detect submarines. This led to sonar (sound navigation and ranging), the World War II anti-submarine technique that uses ultrasound frequencies above 20,000 hertz, past human hearing. The Scottish obstetrician and gynaecologist Ian Donald learned about sonar during his wartime military service, and in 1958 found ovarian cysts in women using industrial ultrasound units designed to detect flaws in metal. Unlike X-rays, ultrasound does not harm cells and directly images soft tissue. It is now routinely used to examine foetuses in pregnant women (Figure 11) and to image bodily organs, even in real time with the echocardiography technique that displays a beating heart.

Another physics-based technique, MRI, grew out of nuclear magnetic resonance (NMR) spectroscopy, invented in the 1930s by the Nobel Laureate American physicist Isidor Rabi. NMR uses

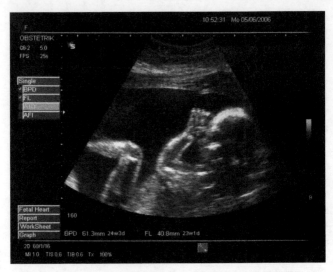

11. Techniques of medical imaging, as in this foetus pictured by ultrasound, depend on applied physics.

the fact that the protons and neutrons in atomic nuclei act like tiny compass needles with north and south poles. If the nuclei are put in a magnetic field, they occupy quantum levels and absorb electromagnetic energy at frequencies that match those levels. The absorption pattern is peculiar to the atoms or molecules involved, making NMR valuable to identify chemical species and study molecular structure.

Then other researchers found ways to observe NMR in biological systems, which always contain water molecules with hydrogen atoms whose single proton nucleus can be detected by NMR. This turned NMR into the medical imaging method of MRI, in which a computer converts the measurements into images of the body's neural, cardiovascular, or musculoskeletal systems and their organs. The method requires intense magnetic fields that come from superconducting metal coils typically cooled with liquid helium to 4 kelvin, just above absolute zero.

MRI has been extended into functional magnetic resonance imaging (fMRI), a non-invasive probe of the living brain. In 1990, the Japanese-born biophysicist Seiji Ogawa working at Bell Labs established the basis for fMRI. He showed that MRI is sensitive to the oxygenation level of blood, which changes as the blood flows through areas of the brain where neurons are active. Hence fMRI maps out the parts of the brain that are actually creating mental processes. It has become an important tool for neuroscience and brain research, and to diagnose brain conditions such as epilepsy and stroke.

Exploring biomolecules and living cells

The use of fMRI in neural research illustrates that besides providing clinical tools, physical methods can explore living systems, one reason biophysics is flourishing today.

Biophysics began with the discovery of 'animal electricity' by Luigi Galvani in 1780, and biological research was reported at the International Congress of Physics in 1900. The 19th-century German scientist Hermann von Helmholtz was a biophysics pioneer whose work combines physiology, physics, and mathematics, and exemplifies the deeply interdisciplinary and diverse nature of the field. His important work on conservation of energy arose from his study of muscle movement; he examined the difference between objective measurements of physical stimuli such as sound waves and how we humans perceive them; he measured the speed at which nerve impulses travel; and in 1851 he invented the ophthalmoscope, the instrument still used to examine the inside of the human eye.

Physicists' thinking also contributed to fundamental biological science when Erwin Schrödinger, whose equation is essential for quantum theory, wrote *What is Life? The Physical Aspect of the Living Cell* (1944). In it he discussed the role of thermodynamics and 'order from disorder' in living things and proposed that a molecular mechanism could transmit hereditary information. The details of the process, however, were not understood until 1953, when the double helix structure of DNA was discovered. Watson and Crick credited Schrödinger's book with inspiring their search.

Biophysicists now use advanced physical tools such as various kinds of microscopy, lasers, spectroscopy, and imaging—some of which I've discussed—to examine biological systems and living things. Researchers can also now manipulate basic biological units with atomic force microscopy, where a tiny probe physically prods a cell, or with optical tweezers, where a tightly focused laser beam holds and moves individual biomolecules and cells. These tools have given insights, for example, about the relation between the three-dimensional shape and function in protein molecules, and about the properties of normal vs cancerous cells.

Along with these research tools, modelling and analysis of complex biological systems is a growing area that draws on basic physics ideas like thermodynamics and statistical mechanics. Such models are useful for instance to simulate and analyse the brain's neural circuits, and electrical theory is important to model the behaviour of individual neurons. Even quantum effects are thought to have a role in some biological processes such as photosynthesis.

The diversity of biophysics and its overlap with other scientific areas like biochemistry and cell and molecular biology makes it difficult to summarize all its research achievements and possibilities. Physical tools, analysis, and theory will play a growing and crucial role in biological science, but it is too early to say if these approaches will carry through to account for the whole range of biological scale, from molecules to cells, organs, organisms, and eco-systems, or to develop whole new theories of biological function.

Clean energy

Physics contributes to human health and longevity through what it offers to biomedicine. It also contributes as we weigh the growing global need for energy against the environmental and health costs of its production and use. Physics is a big part of understanding and developing energy and its sources, from thermodynamics to the principles of electric power generation and nuclear fission, along with nuclear fusion, solar power, and energy from wind, waves, and geothermal effects. Physics also offers possibilities for the more efficient use of energy whatever its source.

The generation of electricity depends on the principles of electromagnetism discovered by Faraday in the first third of the 19th century, which led to the design of generators that rotate coils of wire in magnetic fields to make electric current. Today in the US and globally about two-thirds of the electricity we use is produced at central power plants that burn fossil fuels—coal, and natural gas—to turn the generators. These burning fuels produce the CO_2 and other greenhouse emissions that seriously contribute to climate change and pollution.

One alternative comes from nuclear physics. The discovery in 1938 of nuclear fission in uranium with its huge energy release led to the construction of the atomic bombs that ended World War II. After the war, new efforts turned to the use of controlled nuclear power to make electricity. The first commercial nuclear reactor began operating in 1960, and as of 2014, about 11 per cent of the world's electricity came from nuclear reactors. In one sense this is environmentally favourable. Nuclear plants produce little in the way of greenhouse gases compared to fossil fuels, but have undesirable aspects as well. The nuclear accidents at Three Mile Island, Chernobyl, and Fukushima Daiichi in 1979, 1986, and 2011, respectively, show that nuclear reactors can have serious safety issues. They also produce nuclear waste, byproducts that remain harmfully radioactive for long periods and cannot easily be contained or discarded.

Nuclear fusion, the process that makes our sun and other stars glow as hydrogen nuclei combine to form helium and release energy, offers another possibility. It is inherently less radioactive and safer than nuclear fission but it requires temperatures of many millions of degrees to force nuclei to merge. After decades of research, managing the resulting hot plasma remains a formidable challenge. The International Thermonuclear Experimental Reactor (ITER), being jointly built in southern France by thirty-five nations, is the latest ambitious attempt to control the high temperature plasma with magnetic fields. Another approach at the National Ignition Facility (NIF), Lawrence Livermore Laboratory, California, attempts to initiate fusion with huge lasers.

Solar power could provide a completely different approach to clean and sustainable energy. In passive form, it means designing structures to make maximum use of the energy in sunlight; in active form, sunlight generates electricity. One technique concentrates sunlight with mirrors or lenses and converts it into heat, which drives a steam turbine that drives an electrical generator. A more direct conversion method depends on applied condensed matter physics and materials science. It uses solar cells made of photovoltaic semiconductors, materials whose electrons ordinarily occupy energy levels where they are immobile. But if the electrons gain energy from incoming photons, they make a quantum jump across a 'band gap' to a higher level where they flow as electric current.

The first working solar cell, made from semiconducting silicon, was shown at Bell Laboratories in 1954. Though useful for special applications such as powering space satellites, early solar cells were too expensive and inefficient for general use. But further research on their physics and their semiconducting materials has made them better and cheaper. Their efficiency in converting light to electricity has increased from 10 per cent in the 1970s to a world's record 46 per cent for certain types of cells, demonstrated in 2014; and the cost to make 1 watt of solar electrical power has dropped from US$100 in the 1970s to about US$1 as of 2015. The use of solar electricity is growing but it still provides only a small part of total global electrical consumption and has yet to become a major factor in clean energy.

The physics of matter can provide another big improvement in energy use, the replacement of conventional light sources with efficient light emitting diodes (LEDs). An LED is a semiconductor structure that glows when a small voltage boosts electrons across the band gap to higher energies, after which they drop back down and release photons. The colour of the light depends on the particular semiconductor; in early LEDs it was only red, green, or yellow. It took years of research culminating in a Nobel Prize in 2014 for the Japanese physicists and materials scientists Isamu Akasaki and Hiroshi Amano and the Japanese-born American engineer Shuji Nakamura to produce a blue-emitting LED based on the semiconductor gallium nitride (GaN). With the colour blue now available, LEDs can produce the mixture of colours people consider desirable 'white' light for general illumination.

These new light sources produce more light per watt of electrical power input than incandescent or fluorescent lamps, develop little waste heat and last ten to a hundred times longer. Since lighting takes up a quarter or more of the world's electricity production, white LEDs can provide important energy savings. Another advantage is that their low operational voltages can come from solar cells, making lighting available for the 1.5 billion people worldwide without access to electric power grids. One prediction is that LED lighting will make up some 60 per cent of the global lighting market by 2020.

Kitchen physics

LEDs are not the only way physics enters the home and everyday life. Applied physics has influenced kitchen activities we take for granted, starting with cooking.

The human use of fire goes back 800,000 years to our pre-human ancestors and people have cooked food over open flames for 30,000 years or more. Only in 1800 did cooking advance from this primitive state when Count Rumford, the physicist who concluded that heat is not a substance, invented the kitchen range with a flat cooking surface and an enclosed oven. Early ranges used coal or wood, then gas that burned more cleanly. Finally ranges became flameless when cooking by electric heating coils became widely accepted in the 1930s. A more advanced form of flameless cooking appeared in 1947 when the first microwave oven became available, using the power in microwaves to heat food internally. Fittingly named the Radarange, its technology came from the World War II development of microwave radar.

Cooking over open flames or in inefficient stoves that typically burn biomass is still done by two to three billion people globally, mostly in the third world. The smoke from these fires adds to global pollution and contributes to respiratory problems and other widespread health hazards. Ecologically-minded physicists have used thermal principles such as convection to develop alternative clean-burning yet simple and inexpensive stoves for use in these regions.

Applied thermal physics is essential as well for that other staple of kitchen technology, the refrigerator. Beginning in the 18th century, researchers including Ben Franklin and Michael Faraday confirmed the thermodynamic principle that evaporation of a volatile liquid causes cooling. Others developed the cycle of liquid evaporation and condensation that makes practical refrigeration possible. By the end of the 19th century, these ideas were being put to use in mechanical refrigeration for commercial food preservation. In 1911 the General Electric Company designed a gas-powered refrigerator for home use and in 1927 produced its Monitor Top model, the first electric home refrigerator.

These essentials for kitchen technology rely on 19th-century physics, but increasingly the technology that surrounds us is connected to the modern physics that arose in the 20th century.

Applied modern physics

Though quantum physics and relativity describe parts of nature remote from the human scale, their applications affect our world and our day-to-day lives. Einstein's special relativity produced the equation $E = mc^2$, representing the release of energy at an unprecedented scale in nuclear weapons and nuclear power; as I explained earlier, general relativity enters into the global positioning system; and quantum physics lies behind many of the devices we use daily.

One of these is the laser, which stands for 'light amplification by stimulated emission of radiation'. This device seemed like a gadget out of science fiction and was called a 'death ray' when the first one was built in 1960 by the American physicist Theodore Maiman, based on work by two other American physicists, Charles Townes and Arthur Schawlow. Now lasers have broadly entered our lives in uses from consumer electronics and fibre-optic communications to varied applications in medicine, research, industry, entertainment, and military technology.

The operation of a laser depends on quantum transitions between energy levels that release photons. These can be the atomic or molecular levels found in particular solids such as aluminum oxide (Al_2O_3, used in the first laser), in gases such as CO_2, or levels above and below the band gap in semiconductors. All these media can be made to 'lase', that is, produce photons at specific wavelengths and high intensities that are all in step with each other—the properties that separate lasers from conventional light sources. The result is a range of lasers such as football-stadium size units at the NIF, table-top CO_2 units emitting infrared light at 10.6 micrometres, and tiny semiconductor units emitting in the visible to the infrared for optical fibre communications and Blu-ray optical disc technology.

Quantum physics also underlies the digital devices that increasingly dominate our world. They ultimately depend on the invention of the transistor in 1947 by the American physicists John Bardeen, William Shockley, and Walter Brattain at Bell Laboratories, who shared a Nobel Prize for their work. Transistors utterly changed electronic technology from its dependence on fragile vacuum tubes to the use of these robust solid state units made of semiconducting materials. The quantum mechanical band gap in semiconductors changes their electrical behaviour and makes it possible to precisely control how they conduct current when fashioned into transistors. These highly flexible devices can operate as electronic amplifiers, binary computer bits, and more.

First made as separate units, transistors were soon incorporated into integrated circuits or chips, small flat pieces of semiconducting material, usually silicon, that have been processed to contain millions of transistors and other electronic elements in a tiny area. This technology revolutionized electronics in terms of miniaturization, speed, and power consumption as well as cost of production. It opened up the era of desktop computers, then battery-powered laptops, smart phones, and all the other portable personal devices we now use.

Quantum mechanics is also essential for the new, rapidly growing areas of nanoscience and nanotechnology. The importance of manipulating matter at the smallest scale was expressed in a famous presentation 'There's Plenty of Room at the Bottom' by Nobel Laureate Richard Feynman in 1959. The integrated circuit, developed around that time, was a step in this direction. With the addition of new tools for the small scale such as the scanning tunnelling microscope, there has been an explosion of research in extremely small systems and their uses. This has been supported in the US with a National Nanotechnology Initiative that has provided US$25 billion in Federal research funding since 2001. The European Union and Japan are making other large investments.

Nanoscience focuses on objects with at least one dimension of length 1 to 100 nanometres, or about ten to 1,000 hydrogen atoms in a row. The objects include nanomachines, molecules that perform mechanical operations such as rotation and are expected to have medical applications; and nanoparticles, typically made of gold and other metals or semiconductors. At this scale, quantum effects make nanoparticles function differently from bulk materials, with properties such as their optical behaviour that depend on their size. The resulting versatility supports varied applications in technological devices and biomedicine.

Quantum strangeness

The quantum energy levels that make nanoparticles behave as they do, and also appear in atoms, computer chips, and LEDs, were counter-intuitive when first proposed over a century ago but now are fully accepted parts of nature. Other aspects of quantum physics remain difficult to understand on a visceral level yet these too show up in real applications.

One of these effects is superposition, the fact that a quantum entity like an electron or photon simultaneously represents multiple values rather than a single known value for each of its physical parameters. This comes from the statistical nature of quantum theory, which gives only probabilities. Any parameter value within a certain range is possible until an actual measurement determines a specific value. This non-classical behaviour has been immortalized in the allegory of Schrödinger's Cat, in which a cat placed in a box with a poison flask that may or may not be released by a random trigger is both dead and alive until the box is finally opened. Only then is the cat definitively found in one of the two conditions.

For a real example, consider that the electric field of a photon can be polarized, that is made to point in either a horizontal (H) or vertical (V) direction. H and V could also be labelled 0 and 1 so a polarized photon is also a binary computer bit, but better. An ordinary transistor computer bit is either off (0) or on (1), but the photon is simultaneously 0 and 1. Two bits can hold only one of the binary numbers 00, 01, 10, and 11 (decimal 0, 1, 2, and 3) whereas two quantum bits (or 'qubits') can hold all four numbers. This gives a quantum computer a huge advantage that grows rapidly with the number of qubits. Just ten qubits could hold 1,024 numbers and a hundred-qubit computer could match the power of all the world's conventional supercomputers, at least for certain types of problems such as handling encrypted data.

However, expressing these possibilities in actual hardware is difficult. A qubit could be physically realized with different kinds of quantum particles and systems besides photons, such as electrons that can spin in either of two directions. But there are problems with all of these, a main one being the need to keep the qubits isolated so their superposition is not lost when they interact outside the computer. One of the bigger quantum computers built so far is a twenty-qubit unit constructed by IBM (International Business Machines) for commercial and academic use.

National security and more

With its potential for cryptographic analysis, quantum computation has implications for national security as do many other physics applications. Since World War II, physics has played a major role in advanced weaponry such as nuclear-tipped ballistic missiles and defences against them, smart munitions, and stealth aircraft and ships. Other physics applications for warfare, security, and counter-terrorism include remote sensing from satellites, and underwater sound detection of submarines that can deliver nuclear missiles. Overall, modern military control and communication depends on digital computation and electronics.

In these and other ways, physics and physicists are deeply connected to defence, military applications, and national security across the international scene. In the US, one measure of this involvement is the 2017 DoD budget of about US\$13 billion for basic and applied research and technology development, much of which supports physical science. Other funding comes from the National Nuclear Security Administration. This agency within the US Department of Energy (DoE) is responsible for 'enhancing national security through the military application of nuclear science' and oversees the US stockpile of thousands of nuclear warheads. The agency was supported by a budget of nearly US\$9 billion in 2017.

The role of physics in nuclear weaponry has so far been the single most significant effect physics has had on the US and the world. That evaluation may change with future outcomes; a breakthrough in fusion power, for instance, could have powerful long-term implications for humanity. Meanwhile other applications and interdisciplinary connections affect society as well, over a range from the profound to the entertaining that illustrates the wide influence physics exerts.

Chapter 5
A force in society

Physics is intellectually significant for humanity because of its success in explaining nature, and practically significant because it powerfully affects the wider world outside the laboratory. Physicists recognize this important interaction. The German Physical Society with its over 60,000 members fulfils its 'social commitment' by studying and commenting on sociopolitical issues. At the APS, over 11,000 members of forums called Physics and Society, Outreach and Engaging the Public, and History of Physics consider physics in society. As the Forum on Physics and Society points out on its website:

> Physics is a major component of many of society's difficult issues: nuclear arms and their proliferation, energy shortages and energy impacts, climate change and technical innovation.

Other efforts such as the MIT Program on Science, Technology and Society examine the intersection between society and science, including physics, from the viewpoint of the humanities and social sciences.

Beyond the lab

One powerful contemporary impact of physics began with its role in the Manhattan Project that the US created in World War II to build the atomic bomb—a feat that, as Daniel Kevles writes in *The Physicists: The History of a Scientific Community in Modern America*, came from 'the generation of American physicists who changed the world by forging nuclear weapons'. Those weapons won the war against Japan at the cost of great destruction and were followed by more destructive hydrogen bombs. The possibility of worldwide nuclear devastation was recognized when former Manhattan Project physicists founded the journal *Bulletin of the Atomic Scientists*. Its Doomsday Clock, which predicts the potential for nuclear apocalypse that threatens the human race, is set at two minutes to midnight, reflecting the current global uneasiness about nuclear confrontation (and also about climate change).

Yet contrasting with such a cataclysmic outcome, as the Forum on Physics and Society notes, physics can widely and positively affect our day-to-day activities and the quality of life through its applications in such areas as solar energy, digital electronics, and medicine.

These events and inventions, both fearsome and benign, have appeared within the last fifty to a hundred years, but the societal influence of physics goes much further back and comes from pure as well as applied physics. Besides supporting technology and its innovations, physics responds to the human yearning for answers to deep and long-standing questions.

Origins and our place in creation

One landmark response came when the Copernican revolution changed our view of ourselves. After Copernicus, Brahe, Kepler, and Galileo observed nature, analysed data, and concluded that the planets and stars do not revolve around a fixed Earth, humanity had to reconsider its supposed centrality in the cosmos. In 1632, when Church officials found that Galileo's views conflicted with the official belief in a fixed Earth, they forced him to recant his ideas. It is said that nevertheless he expressed his true belief by rebelliously saying 'and still it moves' but this story is almost surely apocryphal; yet with or without those words, he helped establish the essential principle that human reason can explain the natural world and our position in it.

Other evidence from physics and allied sciences challenged and still challenges religious views about the history of the Earth, the universe, and the human race. I wrote earlier that scientific evidence about the age of the Earth began mounting in the 19th century. Today we have a consensus value of 4.54 billion years while astrophysics gives the universe's age as 13.8 billion years, and sheds light on its origin and development through the Big Bang theory. These results contrast with a literal reading of the Bible, which would imply that God created the universe and everything in it 6,000 to 10,000 years ago. That short time span also does not square with the evidence that humanity and all living things have evolved over several billion years, the overwhelming consensus theory in biological science that is consistent with the age of the Earth.

These results illustrate how physics and science give a new perception of our planet and ourselves at variance with religious ideas and that now dominates in most countries though not the US. For instance, though the scientific consensus for the Big Bang theory is abundantly clear, polls of US adults in 2012, 2014, and 2015 show that a majority do not believe in the theory, or think that even scientists do not fully accept it.

Likewise, the theory of evolution is widely accepted, but not in the US. Polling data in 2006 showed that only 7 per cent to 15 per cent of adults in nine European nations believe that evolution is absolutely false, but that figure was 33 per cent in the US. In 2017, a Gallup poll showed that 76 per cent of US adults either believe that God created humans just as they are now within the last 10,000 years, the so-called creationist or 'intelligent design' view;

or that humanity has evolved but with God's guidance. However, the 19 per cent of US adults who today believe in evolution with no intervention by God is double the figure from 1982, and the number is greater among the more highly educated. Apparently scientific findings do change the general consciousness, if slowly.

But though there is strong physical evidence for how the universe was born and has developed, physics cannot yet examine conditions before or at the exact instant of the Big Bang to fully answer the question 'How did the universe come to be?' with a scientific narrative.

The technological physicist

More pragmatically, physics has changed how we live and work through its effects on the material conditions of human society, especially since the Industrial Revolution. I have already mentioned how the inventions and processes at the heart of that revolution, beginning with the steam engine, both drew on and contributed to classical physics; and how physics principles applied to human needs—that is, science creating technology—changed our cooking habits and much else.

By the second half of the 19th century, physics was having a wide impact through its support of industrial technology. One early example came after the Scottish mathematical physicist William Thomson, Lord Kelvin, developed the electrical theory of telegraphy. He became the scientific adviser to the Atlantic Telegraph Company in its efforts to lay a transatlantic telegraph cable and participated aboard the ships that carried out the work. Initial attempts failed, but Thomson's insights played a big part in the company's final success when it laid nearly 2,100 miles of cable between Ireland and Newfoundland in 1866. This great technological achievement earned Thomson a knighthood from Queen Victoria.

Later, the use of physics to support technology was made a national goal. In 1887, the German industrialist Ernst Werner von Siemens founded the Physikalisch-Technische Reichsanstalt (Imperial Institute of Physics and Technology) to carry out research for industry, which the German government later took over. It was there in 1899 that accurate blackbody spectra were measured for German lighting companies, which led Max Planck to the idea of the quantum. The National Physical Laboratory in the UK and the Bureau of Standards (now the National Institute of Standards and Technology) in the US were established in 1900 and 1901, respectively, on the model of the German institute.

Corporate laboratories arose in that era as well. The first in the US was the General Electric Research Laboratory, founded in 1900 by a group including Thomas Edison. It was described as a 'research laboratory for commercial applications of new principles, and even for the discovery of those principles'. In the Netherlands the Philips Company established a laboratory in 1914, and in the US, Bell Telephone Laboratories, founded in 1925 with 4,000 scientists and engineers, would later become the birthplace of the transistor. Research at another technology-oriented company, IBM, has roots that go back to 1945. This research produced the Nobel-Prize winning discovery of 'high temperature' superconductors that operate far above absolute zero though still at extremely cold temperatures.

Today physicists at corporate facilities play a substantial role in the private sector. A 2014 report from the US Congressional Research Service counts some 274,000 physical scientists within the national workforce of scientists and engineers in 2012. Among the physicists included in these, roughly half are employed in the private sector—not only those with Bachelor's and Master's degrees, but highly trained research-oriented PhD physicists—according to a 2015 report from the American Institute of Physics.

Most of these PhDs contribute to physics, engineering, computers, and other scientific or technological fields. Surprisingly, 7 per cent of them work in an area that seems remote from physics, the world of finance or 'econophysics'. They are quantitative analysts or 'quants' who use their training in mathematics, data analysis, and modelling for investment and commercial banks, hedge funds, and portfolio management companies; or who develop software to make extremely fast automated buy and sell decisions in the stock market. These exotic approaches have enriched some people, but others think the opacity and complexity of the algorithms contributed to the global financial meltdown of 2008 and to market volatility. The role of the quants in the financial industry may have widespread effects we have yet to fully understand or control.

Physics also affects society by inspiring entrepreneurs who develop innovative and potentially commercially successful technology based on physics, as shown by the career of Elon Musk. His undergraduate physics training helped him establish the pioneering technologies in his aerospace company SpaceX and electric car company Tesla Motors, each with the possibility to radically change established practice. In a recent interview, Musk said that the study of physics is good preparation for innovation because it teaches how to reason from first principles:

> [I]f you are trying to break new ground and be really innovative, that's where you have to apply first-principle thinking and try to identify the most fundamental truths in any particular arena and you reason up from there.

Nathan Myhrvold provides another example of using physics in an entrepreneurial manner. Trained as a theoretical physicist, after working as Chief Technology Officer for Microsoft he founded The Cooking Lab to pursue a novel approach to cooking originally called molecular gastronomy. He and other practitioners of the style, which Myhrvold calls modernist cuisine, have used physics-based methods to improve how meat is cooked, to better understand how bread is baked, and to create a different version of the famous Baked Alaska dessert. Efforts like Musk's and Myhrvold's are inspiring universities and professional organizations to help physics students and practising researchers add the appropriate mindset and business training for careers as entrepreneurs.

At war

Physics affects society through its roles in the academic world and in government laboratories and agencies as well as the private sector. In all three areas, physics in the US and elsewhere has prospered at least partly because it is crucial to national defence. For good or ill, physics has advanced warfare and warfare has advanced physics, even if only implicitly.

Before there was a true science of physics, medieval weaponry applied simple mechanical principles, such as using a counterweighted arm on a pivot or the energy stored in a twisted rope to launch heavy projectiles. These siege engines were not designed by rigorous physical analysis, but when they were replaced by artillery, accurate calculations of the projectile's path became important. The Italian mathematician Niccolò Fontana Tartaglia analysed the behaviour of cannonballs in 1537 and correctly claimed that the greatest range came with the cannon elevated at 45 degrees. Later the mechanical principles put forth by Galileo and Newton led to a full science of ballistics.

Physics and physics-based technology have entered into warfare more recently, though not always embodied in weapons. The American Civil War saw the use of balloons and improved telescopes for observation, and the telegraph for communication (President Abraham Lincoln commanded his armies direct from the White House) along with rifled artillery, submarines, and iron warships. That war might also have seen the directed application of science for military use after Lincoln established the National Academy of Sciences (NAS) during the conflict in 1863. The NAS was charged with advising the nation about scientific matters, which it still does; but at the time it did not oversee US science or help the North win the Civil War.

However, fifty years later the NAS formed the National Research Council (NRC) to coordinate scientific research for the military during World War I. After the US declared war against Germany in April 1917, the head of the NRC, astrophysicist George Ellery Hale, worked with Michelson and Robert Millikan (a Nobel Laureate and future Laureate in physics, respectively) to oversee relevant research in physics (other sciences were represented as well). The physics contribution was mostly in detection methods, such as a project to sense submarines by ultrasonic waves that built on Paul Langevin's work in France.

The NRC lacked its own research facilities and worked with industrial laboratories, universities, and military units such as the US Army Signal Corps. An important outcome of this programme, writes physics historian Johannes-Geert Hagmann, was the 'entanglement of scientific, industrial, and military research' with the consequence that later on '[s]cientific and technological research, including major contributions from physics, became a decisive factor in warfare'. These connections persist today, after having been amply illustrated in World War II and the following Cold War.

There is a long list of offensive and defensive weapons created or perfected with government support for physical science and technology on both sides of World War II: jet aircraft, the V-1 flying bomb, and V-2 rocket, sonar, radar, night vision technology, the proximity fuse that made anti-aircraft fire more effective, and, most important, the atomic bomb.

These weapons required governments to organize and fund scientists and engineers, and research and production facilities, at unprecedented scales. In the US, the MIT Radiation Laboratory (or Rad Lab) employed up to 4,000 people including many PhD physicists, and worked between 1940 and 1945 with the British to develop radar, radio navigation, and other electronic devices. The German V-2 rocket, the first long-range guided ballistic missile, was based on rocketry research in Germany and the US in the 1920s and 1930s. The Nazis produced and launched over 3,000 V-2s against London, Antwerp, and Liège, which are estimated to have killed 9,000 people, along with some 12,000 captive labourers on the project.

The nuclear era

The most physics-intensive and historically influential of these military research efforts was the US project to build an atomic bomb. After German researchers discovered nuclear fission in 1938, scientists realized the potential of an uncontrolled fission chain reaction to produce a stupendously destructive weapon. In 1939, the Hungarian-born American physicist Leo Szilard wrote a letter that was signed by Albert Einstein and sent to US President Franklin D. Roosevelt. It warned of German interest in nuclear weaponry and urged the US to begin its own programme to construct 'extremely powerful bombs of a new type'.

The resulting Manhattan Project was initiated in 1942 under Major General Leslie Groves of the US Army Corps of Engineers (and so-named because some preliminary research was carried out at Columbia University in New York). It grew to employ 130,000 people, including European scientists such as the Italian Nobel Laureate physicist Enrico Fermi. It carried out research and production at thirty sites in the US, UK, and Canada at a cost of US\$2 billion, equivalent to US\$27 billion in 2016 dollars.

The Project had to show that a self-sustaining nuclear chain reaction is possible, which Fermi did in the world's first nuclear reactor, in 1942. It also had to separate quantities of the fissionable isotopes uranium-235 and plutonium-239 from their far more prevalent non-fissile forms, an arduous process that required a variety of approaches; and had to find ways to quickly bring these materials to the 'critical mass' necessary to explode, and put them and the detonating mechanism into bombs. These last steps were completed at the Los Alamos Laboratory near Santa Fe, New Mexico, under the American theoretical physicist J. Robert Oppenheimer.

The first atomic bomb, of the plutonium type, was successfully detonated in the Trinity test at the remote Alamogordo Bombing and Gunnery Range in New Mexico on 16 July 1945. On 6 and 9 August, respectively, the US dropped a uranium bomb on Hiroshima and a plutonium bomb on Nagasaki to end the war with Japan, the first and to date the only uses of nuclear weapons in warfare.

Oppenheimer later related that witnessing the Trinity test made him recall what the god Vishnu says in the Hindu sacred text the *Bhagavad Gita*: 'Now I am become Death, the destroyer of worlds'. The Hiroshima and Nagasaki bombs had explosive powers equal to that of 15 to 20 kilotons of TNT, enough to produce an estimated 200,000 or more casualties, level large portions of the cities, and inspire fear and awe about the new atomic age. But more was to come.

In the US, Fermi and two other émigré physicists, Edward Teller and Stanislaw Ulam, conceived and designed a more powerful thermonuclear weapon based on the fusion of isotopes of hydrogen rather than nuclear fission. The US carried out the first full hydrogen bomb test in 1952 at Eniwetok Atoll in the Pacific Ocean. The resulting explosion was equivalent to that from

10 megatons of TNT, and the US began making thermonuclear weapons in quantity. Three years later the Soviet Union tested its own 1.6 megaton hydrogen bomb, and in 1961 detonated a 50 megaton bomb, the largest ever tested. These weapons, delivered across oceans as warheads on intercontinental ballistic missiles (ICBMs), were the basis for Cold War fears that engagement between the US and the Soviet Union would devastate both nations and even threaten the global climate balance.

Other nations have also developed nuclear weapons. Under a 1968 non-proliferation treaty, the US, Russia as successor to the Soviet Union, the UK, France, and China are the only nations allowed to have them; but India, Pakistan, Israel, and North Korea are also known to have them for a total of some 16,000 nuclear weapons, and Iran has apparently been trying to develop them. As Cold War fears have faded, new fears have arisen that failures of diplomacy, rogue regimes, or terrorists could bring nuclear destruction.

These are the painful parts of the complex legacy of the Manhattan Project. A more positive part is that the Project led to the creation of a network of seventeen national laboratories in the US. These carry out fundamental research in addition to directed government work, often at a unique scale of Big Science that only a government could support.

As Daniel Kevles points out in *The Physicists*, another legacy is that the Manhattan Project made physicists so essential for America's national security that they gained 'the power to influence policy and obtain state resources largely on faith'. That brought physics, especially nuclear and high-energy physics, to new prominence and new prosperity. Some of the glow faded in 1993 when the US House of Representatives voted to discontinue funding the Superconducting Super Collider, a US$11 billion project to build a huge particle accelerator 87 kilometres in circumference in Texas. This cleared the way for the LHC at Europe's CERN to begin operations in 2008 as the world's most powerful elementary particle accelerator.

Fundamental research aside, nuclear science still has great military and geopolitical weight. The legacy of nuclear weapons and of physics itself is embedded in culture, the arts, and the media.

The culture of physics

Physics has had varied cultural impacts that reflect its roles in society. Soon after World War II, the atomic bomb inspired films with themes of nuclear destruction. Some displayed overt or hidden nuclear fears in a science fiction format, perhaps to make those terrible possibilities seem less real. In *The Day the Earth Stood Still* (1951), an alien visitor warns humanity of the dangers of nuclear weapons, and the post-apocalyptic film *Five* (1951) is about the survivors of nuclear war. In *Godzilla* (1954) and *Them!* (1954), radiation from nuclear testing produces mutations, respectively an enormous reptile that rampages through Tokyo and giant ants that threaten humanity.

Later, films about nuclear terrors displayed more subtle drama, emotion, and even black comedy. *On the Beach* (1959) showed humanity hopelessly awaiting its end after a global nuclear exchange. *Hiroshima, Mon Amour* (1959), a classic of French New Wave cinema, began as a documentary that director Alain Resnais turned into the story of a brief affair between a French actress and a Japanese architect set against the backdrop of a shattered Hiroshima. Stanley Kubrick's *Dr Strangelove or: How I Learned to Stop Worrying and Love the Bomb* (1964), and Sidney Lumet's *Fail-Safe* (1964), reflected deep fears of nuclear conflict, the first as satire (Figure 12) and the second seriously. Realistic elements also appeared in *The China Syndrome* (1979) about disaster at a nuclear power plant, as would occur at Three Mile

12. **In Stanley Kubrick's nuclear parody *Dr. Strangelove* (1964), a US Air Force officer rides a hydrogen bomb accidentally dropped on the Soviet Union.**

Island, Chernobyl, and Fukushima; and in *Fat Man and Little Boy* (1989), the fictionalized story of the building of the atomic bomb.

The legacy of the atomic bomb endures in the performing arts too, in the stage play *Copenhagen* by the English writer Michael Frayn. After its premiere at London's National Theatre in 1998, it ran elsewhere for a total of 1,400 performances and won Broadway's Tony award for Best Play in 2000. This surprising theatrical success is a minimalist play with three characters on a bare stage talking about physics and its human implications. The characters, drawn from life, are the Danish physicist Niels Bohr, a founder of quantum mechanics; his wife Margrethe; and Werner Heisenberg, who created the Uncertainty Principle and was involved in the German atomic bomb project. Heisenberg had actually visited Bohr in Copenhagen in 1941. *Copenhagen* is Frayn's imagined reconstruction of their discussion about building an atomic bomb and where this would lead.

The nuclear age received further cultural recognition in 2005, when the San Francisco Opera presented the world premiere of the opera *Doctor Atomic* by the American composer John Adams with libretto by Peter Sellars, which treats the Trinity atomic bomb test with Oppenheimer, Teller, and others as characters.

Physics icons

The atomic bomb is not the only cultural legacy of physics, which has produced iconic intellectual figures for our age. Albert Einstein's fame has only grown since the 1919 reports of the solar eclipse that verified his theory of general relativity. Physicists consider him the greatest physicist of all time along with Newton, and to the general public 'Einstein' is synonymous with 'exceptional genius'. A Google search for 'Einstein' yields over a hundred million results and his name appears in the massive Google Books database far more frequently than that of any other scientist.

The late British theorist Stephen Hawking is another iconic physicist. His popular book *A Brief History of Time: From the Big Bang to Black Holes* (1988) sold over ten million copies. His scientific stature drew on his theoretical insight that 'Hawking radiation' can escape from a black hole, an achievement made more notable in the face of his struggle with amyotrophic lateral sclerosis. This neuromuscular disease left him unable to control almost all his bodily movements and confined him to a wheelchair. But Hawking's mind was intact, making him a symbol of the victory of intellect and will over severe physical disability.

Both physicists are well represented in general culture. Einstein's life inspired the opera *Einstein on the Beach* by the American composer Philip Glass, which has been revived several times since its premiere in 1976; and in the film *I. Q.* (1994), Einstein is played by Walter Matthau as smart, wise, and kindly. Hawking's life and work are shown in the film *The Theory of Everything*

(2014), which led to an Oscar for Eddie Redmayne who portrayed him. Hawking also delighted television viewers by appearing in the series *Star Trek: The Next Generation*, *The Simpsons*, and *The Big Bang Theory*. Einstein's and Hawking's achievements have contributed to the media image of the brainy physicist, in fictional characters like theorist Sheldon Cooper (Jim Parsons) in *The Big Bang Theory*—though along with his brilliance, Cooper displays cartoonish levels of quirkiness and social ineptness that are played for laughs.

Ideas associated with both physicists have become common knowledge as well. Except possibly for Newton's equation $F = ma$, Einstein's $E = mc^2$ is the most famous equation in physics, and the fact that nothing can exceed the speed of light is taken for granted in popular culture; for example, on a T-shirt showing Einstein dressed as a motorcycle cop who cautions us to obey the universal speed limit of 186,000 miles per second. Science fiction stories pay homage to this law by inventing seemingly plausible means for spacecraft to travel faster than light, such as the 'warp drive' in the *Star Trek* series.

Einstein and Hawking are also linked to two other physics ideas that captivate the general public, black holes and quantum physics. It was Einstein's general relativity that predicted black holes (also wormholes, theoretical shortcuts through spacetime that connect two black holes as if there were no distance between them, a major plot element in the film *Interstellar* (2014)). Hawking's scientific reputation rests on his quantum analysis of black holes. Quantum physics itself has become part of popular culture: people are aware of the Heisenberg Uncertainty Principle and the analogy of Schrödinger's Cat.

Some popular efforts attempt to connect quantum physics to other ways of knowing. *The Tao of Physics* (1975) by Fritjof Capra drew parallels between modern physics and aspects of Eastern mysticism. It was an influential best seller, but was criticized by physicists such as Nobel Laureate Leon Lederman. Other efforts bring in New Age ideas that lack scientific rigour or that physicists reject as pseudoscientific 'quantum woo'. For instance, Deepak Chopra's *Quantum Healing* (1989) asserted that quantum phenomena affect human health and well-being, and the film *What the Bleep Do We Know* (2004) misleadingly claimed that quantum physics allows mind control of reality.

How physics is presented in popular culture affects general perceptions of the science and the trust people have in it and, by extension, in all science. In return, physics has affected the media themselves, at least in visual art.

Tools for art

Important trends in visual art have been initiated or enhanced as artists adopt physics-based methods. In 1807, the English physician and researcher William Hyde Wollaston patented the *camera lucida* ('bright chamber' in Latin) as an aid to artists. Using a glass prism cut at specific angles and mounted on a stand, it allowed an artist to trace on paper an exact copy of a scene right side up and under normal lighting, instead of upside down and in dimness as required by the *camera obscura* ('dark chamber'), essentially a pinhole camera. The *camera lucida* in turn inspired William Fox Talbot and Louis Daguerre to develop processes to chemically preserve or 'fix' its images on paper, leading to the birth of photography in the 1840s.

When later a variety of artificial light sources became available, artists incorporated them to paint with light. In the 1960s, the American artist Dan Flavin produced subtle arcs of colour from standard fluorescent lamps. More recently, the American artist James Turell uses the evenness of fluorescent lighting to create apparently solid, three-dimensional volumes of light in his installations, and the Danish artist Olafur Eliasson uses fluorescent lamps in works such as his outdoor installation *Yellow Fog* (2008) in Vienna.

Artists have even more eagerly embraced novel light sources. Soon after the laser was invented in 1960, this new tool to produce pure and directed light was put to aesthetic use in installations such as those at the seminal 1971 'Art and Technology' show at the Los Angeles County Museum of Art. Lasers were also soon generating spectacular light shows at rock concerts and other venues, and now find uses in conserving art and creating accurate reproductions of artworks for archival and display purposes.

The latest innovation for artists is the LED. Unlike incandescent and fluorescent lamps, LEDs can be instantaneously turned on and off by computer and so are suitable for dynamic artworks. One example is the large installation *Multiverse* (2008) by Leo Villareal, an array of over 40,000 white LEDs surrounding a 61-metre moving walkway for visitors to the National Gallery of Art in Washington, DC (Figure 13). The LEDs are programmed to

13. Leo Villareal's *Multiverse* (2008) at the National Gallery of Art, Washington, DC, uses thousands of white LEDs.

develop spectacular ever-changing patterns suggesting that viewers are travelling through space amidst stars and galaxies.

Physics matters

From its aesthetic uses to its roles in warfare, physics and the technology it supports influence our society in a variety of ways, some existentially important such as nuclear weaponry. Beyond these impacts, the mere presence of an accomplished science of physics has a special meaning for society in our present age, when science seems to be losing the trust of ordinary citizens. From denial of human-made global warming and of biological evolution to the anti-vaccination movement and hostility to genetically modified organisms, many scientific areas are seen as less than believable or are under attack.

As a science with ancient roots, a roster of eminent thinkers like Einstein, and a solid theoretical base, and with the ability to provide clear answers to problems in the lab and in society, physics is successful. This is not to say that many of its outcomes, like nuclear weaponry, do not need to be seriously weighed by society—they do; but physics is a science that works. Yet its success does not mean that physics has found all the answers to our questions about nature or even within its own applications.

Chapter 6
Future physics: unanswered questions

Standing near the beginning of the 21st century with the 20th century still in sight, we see parallels to earlier times in physics. Just as when the 19th century became the 20th, today physics can look back at recent breakthroughs at all scales, small to large, and in all its usages, pure to applied, such as the discovery of the predicted Higgs boson and the surprise discovery of dark energy; successes in exploring space and finding exoplanets, and in examining our own planet; achievements in novel electronic and photonic technology, and in biological physics with new tools to probe living systems and ourselves.

These results show that physics continues to deepen our understanding of nature and affect how we live our lives, and raises new questions while addressing old ones. A century ago, the new questions included explicit ones such as 'What is the ether that supports electromagnetic waves?' and 'Are photons real?' Humanity has always asked broader queries as well, such as 'How did the universe begin?', 'How will it end?', and 'Is there life elsewhere?' Physics research in the 20th and 21st centuries has answered those explicit questions and has brought at least the beginnings of answers to the broader ones.

The questions that inspire today's continuing research and its future arise from those results—questions such as 'What lies beyond the Standard Model?', which lead to a better understanding of nature; and others such as 'How can we produce cleaner energy?' whose answers, we hope, will improve the human condition.

Physics in the 21st century

Within the first category physicists have written extensively about the present and future of particle physics, cosmology, and so on. A 2001 report from the NRC, *Physics in a New Era*, offers a wider view. Written by a panel of distinguished scientists, it treats both 'Physics Frontiers', covering big questions such as how the universe evolved; and 'Physics and Society', about the status and future of physics in such areas as biomedicine, energy and the environment, and national security. Using this framework, we can see how physics has progressed since this century began, examine the questions that physicists are currently asking, and consider where physics will go next.

Two unexpected physics findings discussed in the NRC report have become more significant since 2001 as we learn how essential they are in the universe: the discovery of a new entity, 'dark matter', which does not emit or interact with electromagnetic radiation and so is invisible to the eye, infrared light, radio waves, and so on; and the discovery that the universe is expanding at an accelerating rate, later connected to an unknown 'dark energy' that acts like a negative pressure, pushing apart the galaxies making up the universe.

Unlike dark matter and dark energy, two other major physics accomplishments since 2001 were anticipated in theory. One is the discovery of the Higgs boson in 2012, a result predicted by the Standard Model and that validates it, yet also illuminates what it lacks. The other achievement further confirms predictions of another bedrock theory, general relativity, and provides a new tool to explore the universe. That was the first LIGO observation in 2015 of gravitational waves followed by other observations since.

The dark universe

Though dark matter and dark energy are crucial to how the universe works, they remain mysteries despite intensive research. Dark matter was first found indirectly through its gravitational effects. After earlier astronomical data hinted at invisible cosmic matter, in the 1970s the American astronomer Vera Rubin noted anomalies in the dynamics of spiral galaxies such as our own Milky Way. Like any rotating body, these enormous pinwheels would throw off any parts of themselves not held firmly in place—in this case, by gravity from the bulk of the galaxy, which depends on its mass. Rubin found it would take many times the amount of matter actually seen in a particular galaxy to keep the stars at its rim from flying off, giving the first solid sign of dark matter.

Dark energy was also discovered indirectly. In 1998, astronomers found that certain supernovae—enormous explosions of stars in their death throes—billions of light years away were dimmer than expected, which meant they were further from us than the theory of the expanding universe predicted in the aftermath of the Big Bang. It was a great surprise when the new data showed an increasing rate of expansion, contrary to the expectation that the pull of gravity from the mass of the universe could eventually slow or reverse the expansion.

Einstein had considered such a force in 1917 when he added a 'cosmological constant' to his theory of general relativity, a kind of anti-gravity that pushed the galaxies apart to keep the universe from collapsing in on itself. But when Hubble showed that the universe is indeed expanding, Einstein dropped the idea, an act he later called his 'biggest blunder'. The 1998 discovery of the accelerated expansion has revived a notion like the cosmological constant under the name dark energy.

Further strong evidence of a dark universe came in 2003 when NASA's Wilkinson Microwave Anisotropy Probe satellite examined the CMB to probe the early universe, followed by a more precise study of the CMB by the Planck Satellite launched by ESA in 2009. The Planck results show that the cosmic mass-energy constitution is only 5 per cent ordinary matter, compared to 26 per cent dark matter and 69 per cent dark energy that add up to 95 per cent of the universe. These proportions define a cosmological 'standard model' that successfully describes many features of the universe including its accelerating expansion and its distribution of galaxies, and sets its age at 13.8 billion years.

The proportions are also stunning evidence that after much effort, humanity has examined only a tiny part of reality and knows little about the rest of it, such as the source of dark energy. One proposal brings in the quantum scale, to suggest that dark energy is actually the so-called 'vacuum energy' arising from the swarm of virtual elementary particles that randomly appear and disappear throughout space. Vacuum energy would increase as the universe gets bigger, fuelling further expansion at an increasing rate.

This would be an attractive scenario but the numbers do not add up. The vacuum energy is calculated to be 10^{60} times bigger than is needed to explain the original supernovae data that led to the discovery of dark energy. Drawing on the multiverse theory, one proposed explanation of this staggering discrepancy is that our particular universe happens to have an extraordinarily small vacuum energy—though this argument illustrates the difficulty of reaching definite answers when one can simply claim that physics is different in a different universe. Or maybe the universe has again entered a time of rapid expansion, as is thought to have happened soon after the Big Bang, or maybe gravitational theory needs to be modified for immense cosmic distances. At this point, we just do not know.

Dark matter raises its own questions. The Standard Model supposedly describes all the building blocks of matter—quarks; leptons, which include electrons, muons, tau particles, and neutrinos; and the Higgs boson (see Figure 8). None of these has the right properties to form dark matter and so the model needs a new particle. The leading candidate, backed up by some astronomical evidence, is a so-called weakly interacting matter particle (WIMP) that barely affects ordinary matter.

But despite tantalizing hints, thirty years of experiments have been fruitless in finding definitive evidence of WIMPs, including recent results from groups in Italy and China. This has unsettled dark matter research and physicists are considering other possibilities. One is to extend the Standard Model by a 'dark sector', a group of new particles and forces tenuously connected to the known ones, for instance as dark photons and dark Higgs bosons that occasionally appear among their regular counterparts. Another suggestion is that the Higgs boson disintegrates into photons and dark matter particles that can then be detected; but as of summer 2018, measurements at the LHC of the Higgs boson show that it decays into two 'bottom' quarks (see Figure 8), a prediction from the Standard Model, with no signs of dark matter particles.

Deeper into the small, mid-size, and large

Even without the disruptive effects of dark matter and dark energy, the Standard Model of particle physics needs modification. Though it is a big step towards understanding the universe, it fails to describe dark matter and dark energy, and it excludes gravity.

It also has structural issues. Quarks and leptons appear in three 'generations' with the same quantum properties and electric charge but hugely different masses (see Figure 8). Electrons, muons, and tau particles are identical except that their masses differ enormously, by a factor of over 3,000 between electrons and tau particles. The six quarks also divide into two sets of three generations each with wildly different masses; and though the Standard Model gives neutrinos a mass of zero, we have actually found three types of neutrinos that differ in their (very small) masses. Another difficulty is that the measured mass of the Higgs boson is far less than it should be according to the quantum theory behind the Standard Model. We simply cannot explain why these elementary particles have the masses that they do.

Besides this, recent results from CERN seem to show that a particle called the B meson behaves differently than the Standard Model predicts. Hoping that this is only an apparent flaw that may point towards a better theory, the KEK high energy accelerator in Tsukuba, Japan, is carrying out high precision measurements of B meson behaviour. Then there is antimatter, made of antiparticles such as the positron. According to the Big Bang theory, matter and antimatter were created in equal amounts at the birth of the universe, yet today we see little antimatter—a fact the Standard Model does not explain.

These and other cracks in the Standard Model guarantee that the thousands of physicists at CERN and elsewhere will continue to gather data and form new theories. Researchers believe the Standard Model is only an approximation to a deeper theory that applies at far higher energies, which appeared in the early universe but that we have not yet reached in the laboratory. The hope has been that experiments at higher energies will find new particles predicted by supersymmetry, a proposed theory that resolves many of the issues with the Standard Model; but so far no such particles have been found. Many observers think this represents a crisis in high energy physics.

To complement research in particle physics, new astrophysics is coming out of LIGO, based at two sites in the US, and the similar Virgo system in Italy operated by a European consortium. Both use kilometres-long laser arrangements to detect extremely tiny fluctuations in spacetime as gravitational waves from cosmic events travel past the Earth. After LIGO first detected these waves in 2015, it began working with Virgo to implement the new technique of gravitational wave astronomy. This capability was shown in spectacular fashion in autumn 2017 when both systems sensed gravitational waves from the collision of two neutron stars, the dense cores left behind after stars of a certain size undergo supernova explosions.

Unlike the original black hole LIGO event, this collision also generated electromagnetic waves, first detected as gamma rays. Then when the event was accurately located in a particular galaxy in the constellation Hydra, some seventy astronomical observatories around the world observed it through its emission of visible light, X-rays, and radio waves. The dual electromagnetic and gravitational analysis showed that collisions between neutron stars can cause the extremely powerful cosmic gamma ray bursts that have been known for some time; and that such collisions can create gold and other elements heavier than iron that we know on Earth, which had been thought to come from the collapse of a single star. These results and the joint observations that led to them were among the scientific breakthroughs of 2017.

Besides far-off cosmic events, we continue studying our solar system neighbours, as with NASA's Juno spacecraft that is examining the atmospheric dynamics of the planet Jupiter. After the Hubble Telescope, the bigger James Webb Space Telescope will observe the solar system as well as distant cosmic objects after its launch in 2021 (Figure 14). Adding to the discovery of water at various sites in the solar system, in early 2018 a NASA orbiter detected water ice just beneath the surface of Mars, and in summer 2018 ESA scientists reported the radar observation of a lake of salty liquid water 20 kilometres across beneath the Martian south pole—important results for potential human exploration of the planet and for the chances of Martian life.

14. NASA plans to launch the James Webb Space Telescope in 2021.

Space satellites in NASA's Earth Observing System also continue to explore our planet while geophysical methods probe its dynamics and structure. The ROMY installation near Munich, Germany, will use ring lasers, where the light beam is split into two beams that travel in opposite directions around a closed circuit, then recombine. As ROMY rotates in space with the Earth, the beams moving with and against the direction of rotation cover slightly different distances, producing a phase difference that depends on the rotational speed. ROMY will measure tiny changes in the length of the Earth's day, the angle of our planet's spin axis, and the ground motion and internal twisting that seismic events cause. It may also advance relativity theory by measuring the effect of 'frame-dragging' on the Earth's spin, the prediction from general relativity that a rotating body warps nearby spacetime.

What is the quantum and does that matter?

The intimate connection between the structure of the universe and its quantum properties is a reminder that nature really is interconnected through all its levels. For that reason and to better understand and use quantum physics, we continue to study the quantum itself. Though quantum theory has convincingly proven its validity over the last century, it remains a puzzle. In 2011, the Swiss quantum experimentalist Anton Zeilinger and colleagues asked thirty-three physicists, philosophers, and mathematicians sixteen questions about basic quantum ideas such as the randomness in nature that quantum theory apparently supports. None of the multiple choice answers was chosen by all participants and many questions prompted widely varying opinions.

This lack of a unified understanding shows the need for a better basis for quantum physics, which is based on a patchwork of ideas; the Schrödinger equation, for instance, is an intuitive guess that works. Some theorists are searching for ways to put quantum theory on a firmer and clearer basis, starting from axioms that treat the theory as a problem in probability or in the transfer of information between systems. Researchers have indeed derived quantum behaviour from such axioms, but they have not yet produced the 'aha' moment that the American theorist John Wheeler thought we may find 'not by questioning the quantum, but by uncovering that utterly simple idea that demands the quantum'.

Meanwhile new experiments are probing exotic quantum phenomena that we can use even without full understanding. Non-classical quantum superposition is the reason a qubit carries more information than a standard computer bit. Another such phenomenon is entanglement, which means that once two quantum particles such as electrons or photons have interacted, they remain linked. Even if widely separated, when a property of one of the particles is measured, the other immediately responds through empty space, a result Einstein called 'spooky action at a distance'.

Entanglement can be illustrated with two photon qubits, prepared so we know they are oppositely polarized, one horizontal and one vertical (digital 0 and 1). However, we do not know the polarization of each photon because both results are possible until a measurement is actually made and sets a definite value. The instant that photon A (say) is measured, whatever the result, a measurement of photon B shows that it has immediately taken on the other value. This occurs no matter how distant B is—as if quantum information were teleported from A to B through some unknown channel faster than light can travel.

Despite this strange, ill-understood behaviour, we could use entangled qubits to produce faster quantum computation by providing additional paths for data storage and flow. They could also guard against the loss of superposition and hence of data because any error would be immediately detected when it destroys the entanglement.

Applications of such exotic quantum effects are being avidly pursued and are well-supported. In the US, besides the twenty-qubit IBM computer mentioned earlier, the company is planning a fifty-qubit unit for the near future. Google is also working on quantum computation, and among US government agencies, the DoE has committed US$40 million to the technology. Elsewhere the European Commission has announced a €1 billion quantum technology project, and Chinese scientists have been especially active in quantum research.

A group under Jian-Wei Pan, at the University of Science and Technology of China in Hefei, recently simultaneously entangled ten qubits in a prototype quantum processor, and studies entanglement itself. In 2017, the team confirmed that entanglement holds over the greatest distances yet measured. They sent entangled photons between two Earthly locations 1,200 kilometres apart by beaming the photons from the first site up to a space satellite, then down to the second site. The advantage of sending photons through space rather than directly between the two Earth sites through optical fibre or open air is that vacuum offers less interference with their quantum states. A satellite network using entangled photon qubits could provide uniquely secure global communications, since any intervention by a third party would ruin the entanglement, as Pan's team has also demonstrated.

Towards quantum gravity

Satellite measurements could also give a unique opportunity to study quantum effects interacting with gravity as entangled photons traverse the varying gravitational field between the Earth and the satellite. A proposal to ESA for the QUEST (Quantum Entanglement Space Test) project to make such a measurement aboard ISS has just passed a feasibility study and could be carried out by the early 2020s. This is one of the few experiments that could examine quantum mechanics and general relativity simultaneously with current technology to provide clues towards a theory of quantum gravity.

Measurements of entanglement in space take on added importance because of recent results that show a tantalizing link between quantum physics and general relativity, which predicts the phenomenon of the wormhole. No wormholes have ever been found; but in 2013, the Argentine theoretical physicist Juan Maldacena and the American theorist Leonard Susskind showed that the two black holes at the ends of a wormhole behave like entangled quantum objects, suggesting that entanglement can create structures in spacetime. Some theorists take this further, thinking that entanglement may create spacetime itself and through it, gravitation, though the details have yet to be understood.

The right approach to a final theory of quantum gravity may be to combine ideas from gravitational theory with—surprisingly—the idea of the qubit and its role in information transfer. A project called 'It from Qubit' follows this strategy by bringing together leading theorists in these fields in a concentrated effort to develop a theory of quantum gravity.

Any approach that is backed by definitive measurements in space would lay to rest proposals to accept 'post-empirical' theory without experimental confirmation. Then we could confidently enter an era of 'post-modern physics' that uses well-tested tools to bring us to an understanding of quantum gravity and new insights into nature.

Energy challenges

Quantum gravity is the hard problem in pure physics, and practical quantum computing is a hard problem in applied physics, but not the only one. Since the 1950s physicists have tried to generate clean and abundant energy from nuclear fusion. This amounts to creating a controlled artificial star on Earth in which hydrogen nuclei collide and fuse into helium, turning mass into energy according to $E = mc^2$. Since positively charged hydrogen nuclei—protons—repel each other, they can be made to merge only when they gain kinetic energy at temperatures of tens of millions of degrees. No material structure could contain such a hot plasma, so the approach has been to use tokamaks (the name comes from an acronym in Russian)—big doughnut-shaped machines where magnetic fields keep the hot plasma away from the walls that contain it.

No tokamak has yet created an energy surplus or has sustained fusion for more than a short time. The ITER project in France, now building a huge reactor housed in a 60-metre-tall building, aims to become the first magnetic confinement effort to create more energy than is needed to heat the plasma. It proposes to produce 500 megawatts in a stable reaction that uses deuterium and tritium (the hydrogen isotopes with one and two neutrons respectively in the nucleus). But ITER has suffered billions of dollars in cost overruns and is years behind its initial schedule, leading to a change in management and decreased confidence from some member nations. Even if its first goals are met in 2035 as now projected, it does not plan to convert the power it generates into electricity, the next essential step towards a practicable power source.

A second method, inertial confinement, aims to initiate fusion by exposing hydrogen isotopes in a millimetre-size volume to extremely intense radiation from 192 focused lasers at the NIF. After years of effort, the results so far are discouraging. In 2016, a report sponsored by the DoE—which has put over US\$3 billion into the NIF (the funding also supported experiments to gather data for nuclear weapons)—questioned whether the project can achieve fusion at all.

Between scientific uncertainty about making hydrogen fusion practical, and possible erosion of government support after several unsuccessful decades, the outlook for near-term results from ITER and NIF must be considered doubtful. However, other approaches are under consideration. Several private companies have raised funding to develop designs for smaller and cheaper fusion reactors. One collaboration, between a private firm and MIT, plans to produce fusion energy within fifteen years.

Another clean energy source, solar power, is more advanced because the photovoltaic conversion of sunlight into electricity is an established process. According to a recent report in *Science* magazine, photovoltaic cells could eventually provide more energy from the sun than the world currently uses. At present, though there are some large solar power installations, in total they deliver only 3 per cent of the approximately 7 terawatts (1 terawatt = 1,000 gigawatts = 10^{12} watts) of electrical power the world generated on average in 2015 (Figure 15); but the report estimates that with the rapidly growing adoption of solar cells and reasonable projections of higher efficiencies and lower costs, photovoltaics could be producing several terawatts by 2030.

15. A large-scale photovoltaic installation in China.

However the world generates its power, the physics of light and of semiconductors will help save energy as LEDs replace conventional light sources, a process on its way to commercial success. Initially extremely expensive, LED sources dropped 85 per cent in cost from 2008 to 2013. Coupled with cost savings over the long LED lifetimes, this has made these sources highly attractive. In the US, the DoE reports, their use jumped from 3 per cent of general lighting applications in 2014 to 13 per cent in 2016. The DoE projects 50 per cent savings in energy used for lighting in the US by 2027 with global benefits in reduced greenhouse gas emissions. Still, the full answer to our energy needs will surely be to combine clean energy with greater efficiency.

Material culture

Another route to better use of energy is through superconductivity, where certain metals and other materials lose all electrical resistance at low temperatures. When an electric current passes through a superconductor, none of its energy is wasted as heat. That saving has made superconducting coils invaluable to carry high currents in electromagnets that produce the strong magnetic fields needed for MRI and nuclear fusion, and to control the accelerated charged particles in the LHC. In these specialized uses the superconducting magnets are cooled to 4 kelvin with liquid helium.

Superconductors would also be valuable for long distance electrical transmission lines that do not waste power and in powerful electromagnets that support frictionless magnetically levitated trains—but for these large-scale applications, cooling with scarce and expensive liquid helium is out of the question. Even the best of the high temperature superconductors discovered in 1986 needs to be cooled to 133 kelvin (–140 Celsius), still an impractical temperature—though in 2015, another compound was found that became superconducting at 203 kelvin (–70 Celsius) but only at high pressures. Research in these materials continues, seeking a superconductor that functions at ordinary temperatures and pressures.

Other novel materials offer further technological potential. The group called topological materials illustrates the power of mathematics in physical analysis. Topology is the branch of mathematics that analyses spatial properties that are preserved in a shape as it is smoothly deformed; for instance, if you start with a doughnut, no matter how you pull and twist, it displays only one hole as long as you don't tear it or glue it to itself. Connections between this and physics may seem far-fetched, but in the 1980s physicists found they gained new insight into matter when they analysed its quantum behaviour with topological mathematics. This work culminated in the 2016 Nobel Prize in Physics for the theorists David Thouless, Michael Kosterlitz, and Duncan Haldane for their work on 'topological phases of matter'.

Using this approach, researchers have explained puzzling quantum effects in solids and found exotic new materials such as an electrical insulator that does not carry current in its interior but does on its surface; a 'semi-metal' in which electrons move extremely quickly without encountering microscopic obstacles; and a material that passes light in only one direction. These effects are expected to spark smaller and faster electronic components including devices essential for quantum computers, and more efficient optical fibre networks.

Some particular materials are important components of major efforts in nanotechnology. One family is based on the element carbon, which takes on different forms called allotropes when its atoms bond in varied molecular configurations; for instance, graphene, a single two-dimensional layer of carbon atoms arranged in a hexagonal lattice. Graphene's versatile properties have made it the subject of a €1 billion effort funded by the European Commission. Other three-dimensional forms are called buckyballs or fullerenes because they look like the geodesic domes designed by the American architect Buckminster Fuller. The allotropes also include carbon nanotubes, which are graphene sheets rolled up into long hollow cylinders only a few nanometres in diameter, the size of a handful of hydrogen atoms in a row (Figure 16).

As single molecules, carbon nanotubes have unusual properties. They are the strongest and stiffest materials ever made, their electrical behaviour can be varied depending on how the atoms are arranged around the cylinder, and they strongly absorb electromagnetic radiation. Their applications range from strengthening materials to creating electrically conducting plastics and improved batteries, along with optical uses. One company has set carbon nanotubes vertically on a flat surface like a forest of tiny trees, resulting in the 'blackest black' ever made. It absorbs 99.6 per cent of visible light or any electromagnetic radiation falling on it, and can be used to make stealth aircraft invisible to radar. (Also, researchers have recently used electromagnetic theory and photonic technology to devise 'cloaking devices' that

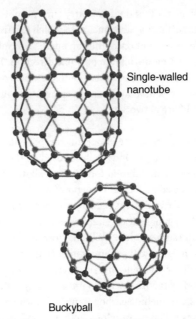

Single-walled
nanotube

Buckyball

**16. Carbon atoms form nanometre-scale cylinders called nanotubes
and geodesic dome-like buckyballs or fullerenes.**

make small objects invisible to ordinary vision, the closest science
has yet come to creating Harry Potter's invisibility cloak.)

Other applications come from semiconductors and metals formed
into nanoparticles typically under 10 nanometres in diameter,
when they display new quantum effects. For example, the
semiconductor cadmium selenide (CdSe) luminesces under
excitation to emit red light at 710 nanometres wavelength.
But when it is fashioned into 'quantum dots', tiny spheres
2 to 8 nanometres across, its band gap changes with size and it
can be made to emit light at 480 to 650 nanometres, blue to red.
Similarly, gold and silver nanoparticles absorb light at specific
wavelengths depending on size. Semiconductor quantum dots are

finding uses in LED and laser light sources, light detectors, and photovoltaic cells. Along with carbon nanotubes and metal nanoparticles, however, their latest and maybe most important long-term possibilities lie in biomedicine and biological physics.

Swallowing the physician

When Richard Feynman made his 'Plenty of Room at the Bottom' speech in 1959, he characterized one potential contribution from nanotechnology as 'swallowing the surgeon'. He meant that someday people would swallow sub-microscopic robots ('nanobots') to perform surgery at specific sites within the body. Such devices are still a long way off, though researchers are developing tiny motors to drive molecule-size machines through a body's blood vessels and organs, and a group in Germany has built a prototype millimetre-size nanobot that can move through a body under magnetic control. We are however closer to 'swallowing the physician', ingesting nanoparticles that diagnose and treat medical conditions without surgery.

In applications combining physics, chemistry, and biomedicine, nanoparticles and hollow nanostructures made of non-reactive gold and silver, non-toxic semiconductors, or carbon nanotubes can carry drugs to specific bodily sites, for instance to treat cancer. Taking advantage of tunable nanoparticle optical properties, efforts are also under way to create 'smart' delivery systems where drug-laden nanoparticles release their cargo when triggered by laser light. In the arena of medical imaging, semiconductor quantum dots can provide a new approach when they are coated to bind to particular proteins or cell types within the body; then, activated by ultraviolet light, they glow at infrared wavelengths that penetrate bodily tissue. When that light reaches external detectors, it provides diagnostic images of internal structures such as tumours. These methods have been successfully tested in mice and are on the verge of clinical testing.

Besides medical uses, nanoparticles give new ways to examine fundamental biological processes. In one example, in 2003 physicist Brahim Lounis at the University of Bordeaux and his colleagues attached a 5-nanometre gold nanoparticle to an individual protein molecule. Then, using a green laser beam to heat the gold and a red laser beam to sense the resulting change in optical properties around the gold, they could follow the dynamics of the protein molecule as it traversed a living cell.

Physics may also provide conceptual frameworks to attack broad problems in biomedicine. One example arises in cancer treatment, where it is thought that physical and network theory could help explain metastasis, the process whereby a cancer moves from a localized tumour to spread throughout a body. Early research in this area is now under way.

Answering ancient questions

We hope and expect that ongoing research in specific areas like dark matter and the Standard Model, or fusion power and new materials, will answer significant questions in physics. Pulling back to a broader vantage point, we see that we are also beginning to answer some of humanity's oldest questions.

The early Greek thinkers, and mythmakers in varied cultures, had ideas about the birth and end of the world. Now the combination of the Standard Model and general relativity, the theories of the small and the large, gives a broad outline to the story of the birth and development of the universe and perhaps its future—though much remains to be filled in such as the nature of dark matter.

According to the Big Bang theory and general relativity, this cosmological tale began in a singularity, a point of infinite density and extremely high temperature. All the physics we know breaks down in such conditions, and it may be that physics will never be able to confirm the exact instant of creation or examine what went before. That lack is one reason that many people find the scientific narrative less compelling than the story of God saying 'Let there be light' and forming the universe; nor does everyone find personal meaning in the physics-based origin story. For those, that meaning will indeed come only from religion or perhaps philosophy; for others, what physics says about the universe and our place in it provides the best path towards meaningful understanding.

The query 'are we alone in the universe?' has also been of enduring interest. We cannot yet give a definitive answer, but since the last years of the 20th century, we have found more water in the solar system than we once thought existed, in unexpected places such as beneath the surface on moons of Jupiter and Saturn, and on Mars; we have discovered other solar systems with possibly Earth-like planets like the seven in the Trappist-1 system, 39 light years away; and in 2018, researchers reported that NASA's Curiosity rover found complex organic compounds by drilling into the Martian surface—compounds that conceivably represent life that once existed there. With this new knowledge, we must conclude that the odds for finding past or present life elsewhere have greatly increased.

We have learned more too about another question that occupied the early Greeks, the constitution of matter. Democritus championed the atomic theory, whereas Aristotle believed the world is made of the four elements Earth, Air, Fire, and Water, and that celestial bodies are made of something different, the Aether. Now modern understanding supports the reductionist atomic theory and provides further understanding down to the level of quarks and up to the level of condensed matter and exotic forms of matter; yet Aristotle was not completely wrong either. We have found an apparently different form of matter by looking to the skies as he did, and have yet to find dark matter in our own earthly sphere.

It is true that Democritus' and Aristotle's ideas were more speculative than what we today call 'scientific', but speculation should still play a meaningful role in the future of physics—for instance, in the desire to find a means to exceed the speed of light as we explore the universe. Yes, the theory of relativity absolutely forbids this for a speeding spacecraft, and though wormholes that would give effective speeds greater than that of light are mathematically possible, we have not found any nor do we see how they could be used in reality.

But thinking and theorizing about such possibilities, always eventually to be confirmed by experiment, should not be left out of the physics imagination. Our experience with serendipity, from the discovery of X-rays to the discoveries of the CMB, the 'butterfly effect', and dark energy, should convince us that we never know how and when the next insight will come.

What was of less interest to the early Greeks now plays a big part in how physics affects the world—I mean the application of its principles to produce technology that people use. I've discussed the yin and yang of physics in society, from the terrors of nuclear war to the peaceful uses of nuclear fusion and solar power, and the benign use of physics in biomedicine. These interactions will only grow, as shown by the numbers of physicists involved in applied and industrial work that relates to technology.

Most of that technology will benefit humanity or at least be neutral in its effects; but physicists should be sensitive to their special and historic role in nuclear weaponry and warfare in general. The biomedical science community is now considering the ethical implications of human genetic engineering even before the technology is fully established. Nuclear weapon technology is already established, but the global physics community would do well to consider its place in a world with new nuclear threats, and how physicists might help to reduce them.

Then there is the future of physics itself as a profession (for many physicists, more than that, a calling). Its character is changing and should continue to do so, especially its heavily male nature. It would greatly benefit physics and society to reach a more even gender ratio so that a woman in a lab is no longer unusual. Driven by social trends and by focused efforts from professional physics societies and other groups in the US, UK, and elsewhere, the number of women in physics is indeed growing, though not quickly. In the US, the percentage of women who earned physics PhDs has increased from 7 per cent to 20 per cent, but that took twenty-nine years, from 1983 to 2012. In 2013, a survey across the European Union showed that women still occupied only 11 per cent of full professor academic positions in the natural sciences and engineering.

In an encouraging sign, however, that same European survey shows that the number of female researchers across all fields has been rising faster than the number of male researchers. Another positive change is that discrimination against, and sexual harassment of, women in the sciences is now taken more seriously, as shown by recent cases at US universities and CERN. These trends suggest that while it will not happen in just a few years, with continuing efforts towards equal treatment, women in physics—and in all science—are on their way to a more equitable position worldwide.

Similar considerations apply to underrepresented racial and ethnic minorities. Their definitions vary by country, but in the US, these constitute African Americans, Hispanic Americans, and Native Americans. Their numbers in physics have grown but still represent under 7 per cent of US PhD physics degrees awarded in 2013–15, whereas these groups formed nearly 30 per cent of the US population in 2010. As is the case for women, new efforts are being made in recruiting more physicists from these ranks, to benefit physics and to fulfil the ideal of the US to be a fully inclusive society—an aspiration that other democratic societies share.

An international enterprise

Another change in physics is its growing internationalization. From its Greek roots, the science of physics grew through different cultures and nations to become international, at least throughout Europe as shown in the 1900 International Congress of Physics held in Paris. The American presence in physics was hugely enhanced by the Manhattan Project during World War II; then, after the war, the US could afford to devote resources to physics and science that other war-damaged nations could not.

Now physics is thriving across the globe. Multi-nation consortiums like CERN, ITER, and ESA carry out major research projects; China and Japan have both proposed building successors to the LHC; and the Indian Space Research Organization plans to send a lander to the moon's unexplored South Pole, and then to examine Mars and Venus. China especially is rapidly becoming a major scientific power. A recent NSF report 'Science & Engineering Indicators' notes that China is now the world's largest spender on research and development after the US. It also produces a prodigious number of research papers, for instance having published about 426,000 articles in 2016 compared to some 409,000 articles from the US.

International physics follows the best traditions of science as a universal human endeavour; but strength in physics research and the technology it supports also has powerful implications for economic and military competitiveness among nations, and for the welfare of their citizens. Changes in where and how physics is done globally could strongly influence future geopolitics and the international order, and the goal of understanding nature cannot be entirely separated from these societal roles—nor should it. The quest, after all, is supported by society, and physics should likewise contribute to society.

The quest

Yet the pure quest at the heart of physics (and all science) has its own deep value apart from political realities, or commercial and military applications. The journey to scientific understanding takes its place alongside art, philosophy, and religion as expressions of the human spirit that seek to make sense of what may seem an uncaring, even fearful, universe.

Our greatest physicists have known this. Isaac Newton has told us that he was like a little boy playing at the seashore 'whilst the great ocean of truth lay all undiscovered before me'. Marie Curie understood the power of knowledge in a famous quote ascribed to her, 'Nothing in life is to be feared, it is only to be understood. Now is the time to understand more, so that we may fear less.' And Albert Einstein reminds us:

> The important thing is not to stop questioning. Curiosity has its own reason for existence. One cannot help but be in awe when he contemplates the mysteries of eternity, of life, of the marvelous structure of reality. It is enough if one tries merely to comprehend a little of this mystery each day.

Future physics will find answers to the newest and the oldest questions, and will aid humanity as well, if it can remember and follow these wise words.

References

Chapter 1: It all began with the Greeks

'To the same degree of refrangibility...', I. Newton, *Newton's Philosophy of Nature: Selections from His Writings*, H. S. Thayer, editor (Mineola, NY: Dover Publications, 2005), p. 74.

'Completed one full turn of the helix of scientific advance...', J. L. Heilbron, *Physics: A Short History* (Oxford: Oxford University Press, 2015), p. 111.

'Age of correlation' and 'cannot possibly be a material substance' are quoted in F. Cajori, *A History of Physics* (Mineola, NY: Dover Publications, 1962), pp. 143, 202.

'While it is never safe to affirm that the future of Physical Science...', A. Michelson, *The University of Chicago Annual Register, July 1895–July 1896* (Chicago: University of Chicago Press, 1896), p. 159.

Chapter 2: What physics covers and what it doesn't

Newton's definitions of absolute space and time are found in I. Newton, *Newton's Principia: The Mathematical Principles of Natural Philosophy*, Andrew Motte, translator (New York: Daniel Adee, 1846), p. 77.

'A wonderful opportunity for YOU...', LHC@home, http://lhcathome.web.cern.ch/about/why-we-need-your-help

The figures cited in the paragraphs beginning 'We can compare what physicists are doing...' and 'Today, judging by the lists...' come from A. K. Wróblewski, 'Physics in 1900', *Acta. Physica Polonica B* **31**, 2 (2000), pp. 179–95; United States Department of Labor, Bureau of Labor Statistics, 'Occupational Employment Statistics',

https://www.bls.gov/oes/current/oes192012.htm; American
Physical Society, 'Official 2018 Yearly Unit Membership Statistics',
https://www.aps.org/membership/units/upload/YearlyUnit18.pdf
(accessed 30 May 2018).

'The physical principles and mechanisms...', American Physical
Society Division of Biological Physics, https://www.aps.org/units/
dbp/biological.cfm. (accessed 17 May 2018).

Chapter 3: How physics works

'The cause and effect must be contiguous...', D. Hume, *A Treatise of
Human Nature* (London: Penguin Books, 1985), p. 223.

The comments about mathematics come from E. P. Wigner, 'The
Unreasonable Effectiveness of Mathematics in the Natural
Sciences', *Comm. Pure Appl. Math.* **13**, 1 (1960), pp. 1–14.

'Even by 1983, only 7 per cent of US physics PhDs...',
Patrick J. Mulvey and Starr Nicholson, American Institute of
Physics, 'Trends in Physics PhDs', https://www.aip.org/sites/
default/files/statistics/graduate/trendsphds-p-12.2.pdf (accessed
30 May 2018).

'We are to admit no more causes...', I. Newton, *Newton's Principia:
The Mathematical Principles of Natural Philosophy*, Andrew
Motte, translator (New York: Daniel Adee, 1846), p. 384.

'I was sitting in a chair...', is quoted in W. Isaacson, *Einstein: His Life
and Universe* (New York: Simon and Schuster, 2008), p. 145.

'Spacetime tells matter...', J. Wheeler, *Geons, Black Holes, and
Quantum Foam* (New York: W. W. Norton, 1998), p. 235.

'In the fields of observation...' is quoted in S. Finger, *Minds Behind
the Brain* (Oxford: Oxford University Press, 2000), p. 309.

The comments about multiverses and string theory can be found in
P. Steinhardt, 'Theories of Anything', *Edge*, '2014: What Scientific
Idea is Ready for Retirement', https://www.edge.org/response-
detail/25405; Richard Dawis interviewed by Richard Marshall,
'String Theory and Post-empiricism', *3:AM Magazine*, http://
www.3ammagazine.com/3am/string-theory-and-post-empiricism/;
G. Ellis and J. Silk, 'Defend the Integrity of Physics', *Nature* **516**,
18/25 December 2014, pp. 321–3; S. Hossenfelder, 'Post-empirical
Science is an Oxymoron', *BackRe(action)*, http://backreaction.
blogspot.com/2014/07/post-empirical-science-is-oxymoron.html
(all websites accessed 18 May 2018).

Comments by Roger Penrose about 'fashionable' physics and string theory can be found in R. Penrose, *The Road to Reality* (London: Vintage, 2005), and R. Penrose, *Fashion, Faith, and Fantasy in the New Physics of the Universe* (Princeton, NJ: Princeton University Press, 2017).

Chapter 4: Physics applied and extended

'The results of your researches…', Nobelprize.org, 'The Nobel Prize in Chemistry 1935: Award Ceremony Speech', http://www.nobelprize.org/nobel_prizes/chemistry/laureates/1935/press.html (accessed 17 May 2018).

Chapter 5: A force in society

'Physics is a major component…', APS Physics: Forum on Physics and Society, 'History of the APS Forum on Physics and Society (1972–2015)', https://www.aps.org/units/fps/history.cfm (accessed 17 May 2018).

'The generation of American physicists…', Daniel J. Kevles, *The Physicists* (New York: Alfred A. Knopf, 1977, 1995), p. ix.

Data about the percentage of Americans who believe in the Big Bang Theory come from Scitable, 'Unbelievable: Why Americans Mistrust Science', Ryan Hopkins, 31 May 2014, https://www.nature.com/scitable/blog/scibytes/unbelievable; *The Atlantic*, 'A Majority of Americans Still Aren't Sure About the Big Bang', Alexis C. Madrigal, https://www.theatlantic.com/technology/archive/2014/04/a-majority-of-americans-question-the-science-of-the-big-bang/360976/, 21 April 2104; Cary Funke and Lee Rainie, Pew Research Center, 'Public and Scientists' Views on Science and Society', http://www.pewinternet.org/2015/01/29/public-and-scientists-views-on-science-and-society/, 20 January 2015 (all websites accessed 30 May 2018).

Data about the percentage of Americans who believe in evolution come from Jon D. Miller, Eugenie C. Scott, and Shinji Okamoto, 'Public Acceptance of Evolution', *Science* **313**, 11, pp. 765–6; and Gallup News, 'In U.S., Belief in Creationist View of Humans at New Low', Art Swift, http://news.gallup.com/poll/210956/belief-creationist-view-humans-new-low.aspx (accessed 30 May 2018).

'Research laboratory for commercial applications...', GE, 'Heritage of Research', https://www.ge.com/about-us/history/research-heritage (accessed 17 May 2018).

'However, fifty years later the NAS formed...', The discussion in this and the next paragraph follows J.-G. Hagmann, 'Mobilizing US Physics in World War I', *Physics Today* **70**, 8 (2017), p. 44. 'Entanglement of scientific, industrial, and military research...' and '[s]cientific and technological research, including major contributions from physics...' appear on p. 50 of that article.

'Now I am become Death, the destroyer of worlds', J. R. Oppenheimer in *The Day After Trinity* (1981), https://www.youtube.com/watch?v=Vm5fCxXnK7Y (accessed 17 May 2018).

'The power to influence policy...', Daniel J. Kevles, *The Physicists* (New York: Alfred A. Knopf, 1977, 1995), p. ix.

Chapter 6: Future physics: unanswered questions

'Not by questioning the quantum...', J. Wheeler, quoted in *Between Quantum and Cosmos*, W. Zurek, A. Van Der Merwe, and W. Miller, editors (Princeton, NJ: Princeton University Press, 2017), p. 12.

Data about women in physics in the European Union comes from European Commission, 'She Figures 2012: Gender in Research and Innovation', http://ec.europa.eu/research/science-society/document_library/pdf_06/she-figures-2012_en.pdf (accessed 23 November 2018).

The figure about US PhD physics degrees awarded to minorities comes from APS Physics, 'Minority Physics Statistics', https://www.aps.org/programs/minorities/resources/statistics.cfm (accessed 30 May 2018). The figure about the percentage of minorities in the US population comes from 'Population of the United States by Race and Hispanic/Latino Origin, Census 2000 and 2010', https://www.infoplease.com/us/race-population/population-united-states-race-and-hispaniclatino-origin-census-2000-and-2010 (accessed 26 July 2018).

'Whilst the great ocean of truth...', I. Newton, quoted in D. Brewster, *Memoirs of the Life, Writings, and Discoveries of Sir Isaac Newton*, Volume II (Edinburgh: Thomas Constable and Company, 1855), p. 407.

'The important thing is not to stop questioning...', A. Einstein, quoted in William Miller, 'Death of a Genius: His Fourth Dimension, Time, Overtakes Einstein', *LIFE* magazine, 2 May 1955, p. 64.

Further reading and viewing

In addition to the works listed below, interested readers will find brief focused treatments of many of the physics topics discussed here in other books in the *Very Short Introduction* series, which cover astrophysics, nuclear physics, quantum theory, relativity, and more.

Chapter 1: It all began with the Greeks

Florian Cajori, *A History of Physics* (New York: Dover Publications, 1962).

J. L. Heilbron, *Physics: A Short History from Quintessence to Quarks* (Oxford: Oxford University Press, 2015).

J. L. Heilbron, *The History of Physics: A Very Short Introduction* (Oxford: Oxford University Press, 2018).

Donald J. Kevles, *The Physicists: The History of a Scientific Community in Modern America* (New York: Vintage Books/Knopf, 1995).

Helge Kragh, *Quantum Generations: A History of Physics in the Twentieth Century* (Princeton, NJ: Princeton University Press, 2002).

National Research Council, *Physics in a New Era: An Overview* (Washington, DC: National Academies Press, 2001).

Chapter 2: What physics covers and what it doesn't

Robert P. Crease, *Philosophy of Physics* (Bristol: IOP Publishing, 2017).

Mathias Frisch, 'Why Things Happen', *Aeon*, 23 June 2015, https://aeon.co/essays/could-we-explain-the-world-without-cause-and-effect (accessed 21 May 2018).

J. Richard Gott and Robert J. Vanderbei, *Sizing Up the Universe: The Cosmos in Perspective* (Washington, DC: National Geographic, 2015).

The Office of Charles and Ray Eames, *Powers of Ten: A Film about the Relative Size of Things in the Universe and the Effect of Adding another Zero* (1977), https://www.youtube.com/watch?v=0fKBhvDjuyO (accessed 21 May 2018).

Caleb Scharf and Ron Miller, *The Zoomable Universe: An Epic Tour Through Cosmic Scale, from Almost Everything to Nearly Nothing* (New York: Farrar, Straus and Giroux/Scientific American, 2017).

A. K. Wróblewski, 'Physics in 1900', *Acta. Physica Polonica B* **31**, 2 (2000), pp. 179–95.

Chapter 3: How physics works

Frank Close, *The Infinity Puzzle: Quantum Field Theory and the Hunt for an Orderly Universe* (New York: Basic Books, 2011).

Walter Isaacson, *Einstein: His Life and Universe* (New York: Simon and Schuster, 2008).

George Johnson, *The Ten Most Beautiful Experiments* (New York: Knopf/Vintage, 2009).

Alan Lightman, *Great Ideas in Physics* (New York: McGraw Hill, 2000).

Royston M. Roberts, *Serendipity: Accidental Discoveries in Science* (New York: Wiley, 1989).

Eugene P. Wigner, 'The Unreasonable Effectiveness of Mathematics in the Natural Sciences', *Comm. Pure Appl. Math.* **13**, 1 (1960), pp. 1–14.

Chapter 4: Physics applied and extended

Philip Ball, *Made to Measure: New Materials for the 21st Century* (Princeton, NJ: Princeton University Press, 1997).

Richard B. Gunderman, *X-Ray Vision: The Evolution of Medical Imaging and Its Human Significance* (Oxford: Oxford University Press, 2012).

Robert Hazen, *The Story of Earth: The First 4.5 Billion Years, from Stardust to Living Planet* (New York: Penguin Group, 2012).

Nano.gov, *National Nanotechnology Initiative*, https://www.nano.gov (accessed 21 May 2018).

John W. Orton, *The Story of Semiconductors* (Oxford: Oxford University Press, 2006).

Michael Riordan and Lillian Hoddeson, *Crystal Fire: The Birth of the Information Age* (New York: W. W. Norton, 1997).

Scientific American, *Understanding Nanotechnology* (New York: Time Warner Book Group, 2002).

Neil deGrasse Tyson, Michael A. Strauss, and J. Richard Gott, *Welcome to the Universe: An Astrophysical Tour* (Princeton, NJ: Princeton University Press, 2016).

Chapter 5: A force in society

Jon H. Else (director), *The Day after Trinity* (film, 1981), https://www.youtube.com/watch?v=Vm5fCxXnK7Y (accessed 22 May 2018).

Johannes-Geert Hagmann, 'Mobilizing US Physics in World War I', *Physics Today* **70**, 8 (2017), pp. 44–50.

Donald J. Kevles, *The Physicists: The History of a Scientific Community in Modern America* (New York: Vintage Books/Knopf, 1995).

Stanley Kramer (director), *On the Beach* (film, 1959), https://www.youtube.com/watch?v=Ue8hC5qqMt4 (accessed 22 May 2018).

Stanley Kubrick (director), *Dr. Strangelove or: How I Learned to Stop Worrying and Love the Bomb* (film, 1964), https://archive.org/details/DRStrangelove_20130616 (accessed 22 May 2018).

Sidney Perkowitz, *Empire of Light: A History of Discovery in Science and Art* (New York: Henry Holt, 1996).

Sidney Perkowitz, *Hollywood Science* (New York: Columbia University Press, 2007).

Sidney Perkowitz, *Slow Light: Invisibility, Teleportation and Other Mysteries of Light* (London: Imperial College Press, 2011).

Richard Rhodes, *The Making of the Atomic Bomb* (New York: Simon & Schuster, 2012).

Michael Riordan, Lillian Hoddeson, and Adrienne W. Kolb, *Tunnel Visions: The Rise and Fall of the Superconducting Super Collider* (Chicago: University of Chicago Press, 2015).

Chapter 6: Future physics: unanswered questions

Steve Fetter, Richard L. Garwin, and Frank von Hippel, 'Nuclear Weapons Dangers and Policy Options', *Physics Today* **71**, 4 (2018), pp. 32–9.

Nancy M. Haegel et al., 'Terawatt-scale Photovoltaics: Trajectories and Challenges', *Science* **356** (6334), 14 April 2017, pp. 141–3.

Catherine Heymans, *The Dark Universe* (Bristol: IOP Publishing, 2017).

Siddarth Koduru Joshi et al., 'Space QUEST Mission Proposal: Experimentally Testing Decoherence Due to Gravity', *New J. Phys.* **20**, June 2018, http://iopscience.iop.org/article/10.1088/1367-2630/aac58b/meta (accessed 10 July 2018).

Mark Miodownik, *Stuff Matters: Exploring the Marvelous Materials That Shape Our Man-Made World* (London: Penguin, 2013).

National Research Council, *Physics in a New Era* (Washington, DC: National Academies Press, 2001).

Sidney Perkowitz, *Slow Light: Invisibility, Teleportation and Other Mysteries of Light* (London: Imperial College Press, 2011).

M. White, *What's Next for Particle Physics?* (Bristol: IOP Publishing, 2017).

Index

SOCIAL MEDIA
Very Short Introduction

Join our community
www.oup.com/vsi

- Join us online at the official Very Short Introductions **Facebook** page.
- Access the thoughts and musings of our authors with our online **blog**.
- Sign up for our monthly **e-newsletter** to receive information on all new titles publishing that month.
- Browse the full range of Very Short Introductions online.
- Read **extracts** from the Introductions for free.
- If you are a teacher or lecturer you can order inspection copies quickly and simply via our website.

CANCER
A Very Short Introduction
Nick James

Cancer research is a major economic activity. There are constant improvements in treatment techniques that result in better cure rates and increased quality and quantity of life for those with the disease, yet stories of breakthroughs in a cure for cancer are often in the media. In this *Very Short Introduction* Nick James, founder of the CancerHelp UK website, examines the trends in diagnosis and treatment of the disease, as well as its economic consequences. Asking what cancer is and what causes it, he considers issues surrounding expensive drug development, what can be done to reduce the risk of developing cancer, and the use of complementary and alternative therapies.

GALAXIES
A Very Short Introduction
John Gribbin

Galaxies are the building blocks of the Universe: standing like islands in space, each is made up of many hundreds of millions of stars in which the chemical elements are made, around which planets form, and where on at least one of those planets intelligent life has emerged. In this *Very Short Introduction*, renowned science writer John Gribbin describes the extraordinary things that astronomers are learning about galaxies, and explains how this can shed light on the origins and structure of the Universe.

www.oup.com/vsi

NUMBERS
A Very Short Introduction
Peter M. Higgins

Numbers are integral to our everyday lives and feature in everything we do. In this *Very Short Introduction* Peter M. Higgins, the renowned mathematics writer unravels the world of numbers; demonstrating its richness, and providing a comprehensive view of the idea of the number. Higgins paints a picture of the number world, considering how the modern number system matured over centuries. Explaining the various number types and showing how they behave, he introduces key concepts such as integers, fractions, real numbers, and imaginary numbers. By approaching the topic in a non-technical way and emphasising the basic principles and interactions of numbers with mathematics and science, Higgins also demonstrates the practical interactions and modern applications, such as encryption of confidential data on the internet.

www.oup.com/vsi